Marine Biology

Structural and Functional Organization of Fish Blood Proteins

MARINE BIOLOGY

Additional books in this series can be found on Nova's website under the Series tab.

Additional E-books in this series can be found on Nova's website under the E-book tab.

MARINE BIOLOGY

STRUCTURAL AND FUNCTIONAL ORGANIZATION OF FISH BLOOD PROTEINS

ALLA MICHAILOVNA ANDREEVA
EDITOR

Nova Science Publishers, Inc.
New York

Copyright © 2012 by Nova Science Publishers, Inc.

All rights reserved. No part of this book may be reproduced, stored in a retrieval system or transmitted in any form or by any means: electronic, electrostatic, magnetic, tape, mechanical photocopying, recording or otherwise without the written permission of the Publisher.

For permission to use material from this book please contact us:
Telephone 631-231-7269; Fax 631-231-8175
Web Site: http://www.novapublishers.com

NOTICE TO THE READER

The Publisher has taken reasonable care in the preparation of this book, but makes no expressed or implied warranty of any kind and assumes no responsibility for any errors or omissions. No liability is assumed for incidental or consequential damages in connection with or arising out of information contained in this book. The Publisher shall not be liable for any special, consequential, or exemplary damages resulting, in whole or in part, from the readers' use of, or reliance upon, this material. Any parts of this book based on government reports are so indicated and copyright is claimed for those parts to the extent applicable to compilations of such works.

Independent verification should be sought for any data, advice or recommendations contained in this book. In addition, no responsibility is assumed by the publisher for any injury and/or damage to persons or property arising from any methods, products, instructions, ideas or otherwise contained in this publication.

This publication is designed to provide accurate and authoritative information with regard to the subject matter covered herein. It is sold with the clear understanding that the Publisher is not engaged in rendering legal or any other professional services. If legal or any other expert assistance is required, the services of a competent person should be sought. FROM A DECLARATION OF PARTICIPANTS JOINTLY ADOPTED BY A COMMITTEE OF THE AMERICAN BAR ASSOCIATION AND A COMMITTEE OF PUBLISHERS.

Additional color graphics may be available in the e-book version of this book.

Library of Congress Cataloging-in-Publication Data
Structural and functional organization of fish blood proteins / editor, Alla Michailovna Andreeva.
   p. cm.
  Includes index.
  ISBN 978-1-62100-264-2 (softcover)
 1. Fishes--Physiology. 2. Blood proteins--Analysis. I. Andreeva, Alla Michailovna.
 QL639.1.S77 2011
 597--dc23
                                                   2011032419

Published by Nova Science Publishers, Inc. † New York

Contents

Preface		vii
Introduction		1
Chapter 1	Principles of the Blood Plasma Proteins Organization in Higher Vertebrates	3
Chapter 2	The Organization of Blood Plasma Proteins in Cartilaginous Fishes *Chondrichthyes*	25
Chapter 3	*Chondrostei* Blood Plasma Protein Organization	41
Chapter 4	Organization of *Teleostei* Blood Plasma Proteins	49
Chapter 5	Distribution of Blood Plasma Proteins between Intravascular and Interstitial Fluids	87
Chapter 6	Structural Conversions of Low-Molecular Blood Plasma Proteins During the Adaptations of Plastic and Water Metabolism in *Teleostei*	101
Chapter 7	The Particular Role of Hemoglobin and Resistance Properties of Erythrocyte Membranes in the Formation of Functional Organization of Fish Blood Proteins	109
Chapter 8	Principles of the Fish Blood Protein Organization	141

Conclusion	153
References	159
Index	181

PREFACE

This book presents a comprehensive analysis of fishblood proteins. It is *Pisces*, in which the highest level of blood proteins structural diversity is revealed, that is why the analysis of their structural organization ways undoubtedly deserves special consideration. The first chapter of the book focuses on the modelsof blood proteins organization and conceptions of blood proteins transcapillary exchange in mammals; the following chapters deal with various aspects of proteins structural-functional diversity in cartilaginous Chondrichthyes and bony Osteichthyes fishes, inhabiting the seas, fresh and brackish waters.

INTRODUCTION

Evolution of the fish covers several geological epochs. After surviving the global climate changes on the Earth, fishes occupied the dominant position in the World Ocean, exceeding all other vertebrates by the number of species. In many respect such success is ensured by the effective mechanisms of stabilization of the internal fluid environment, into composition of which the blood plasma proteins enter.

Modern conceptions about the plasma proteins and their transcapillary exchange in higher vertebrates are based on the model of large proteins, composed of one polypeptide chain, or so-called monomeric proteins, which can penetrate through the capillary wall into the interstitial space in some parts of the capillary network. There are no models of the blood proteins' organization in the fishes. Meanwhile, the osmotic interaction between the fish blood and aquatic environment, the level of salinityof which can fluctuate considerably, require the presence of mechanisms for fast stabilization of the water metabolism, suggesting special features both in blood proteins structure and in their transcapillary exchange. The plastic metabolism in poikilothermal organisms also has distinctive features in blood plasma proteins transcapillary transport. By this reason the fish proteins organization was considered in the close connection with the microcirculation system.

At present, a comprehensive material on the vertebrate blood proteins has been accumulated. The fish proteins occupy the special position because of their supreme structural diversity. To the major degree it accounts for the osmotically active proteins – albumins, among which both simple proteins and glycoproteids, monomers and oligomers with various molecular weights, are found in fishes. Unlike albumins, the globulins organization didn't undergo substantial changes during vertebrate evolution, probably by the reason of the narrow specificity of globulin functions and by the selection trend to preserve these functions. Therefore, analyzing the fish blood proteins from different taxa and biotopes, the special attention has been paid to the low-molecular fraction.

When analyzing the functional organization of fish proteins not only the specific function of some proteins was estimated but the capability of plasma proteins to bind nonspecifically some ligands. The natural proteins usually specialize on binding just some particular kinds of small molecules (except antibodies), and there is a trend in the evolution of proteins to decrease the number of ligands the protein binds (Shulz, Schirmer, 1979). The special consideration in the chapter, which is devoted to the functional organization of the blood proteins, was given to the intracellular protein hemoglobin, which doesn't belong to plasma proteins. Such approach is conditioned by the fact, that hemoglobin molecule stability to degradation plays a major role in developing of plasma proteins ability to bind iron-containing ligands – the products of the destruction of hemoglobin, which enters the bloodstream in result of intravascular hemolysis of erythrocytes.

The book consist of 8 chapters. Chapter 1 describes the conceptions about blood proteins organization and theoretical models of blood proteins transcapillary exchange in mammals. Analysis of the blood proteins structural-functional diversity in cartilaginous*Chondrichthyes* and bony *Osteichthyes* fishes, inhabiting the seas, fresh and brackish waters, is given in chapters 2-4. The factors of proteins diversity formation – external environmental factor (salinity) and organism's internal environmental factor (features of internal fluids composition and of hemoglobin organization), which determined the organization of blood proteins by the monomer/oligomer manner and their specialization level are considered in chapters 7 and 8. The mechanisms of transcapillary exchange of blood plasma proteins and their contribution to stabilization of interstitial fluid filtration processes in the organism are described in chapters 5 and 6. A particular attention is given to the emergence during the evolution of *Pisces* of the dynamically rearranging *in situ* protein blood systems in limnetic *Teleostei*, in which the oligomeric proteins capable of structural transformations are found.

The approaches applied by the author make it possible to reduce the large amount of various forms of fish blood plasma proteins organizations to several discrete types and to distinguish the main strategies of protein blood systems organization in the evolution of *Pisces*.

The study research is supported by the Russian Foundation for BasicResearch, grant no 10-04-00954-a.

Chapter 1

PRINCIPLES OF THE BLOOD PLASMA PROTEINS ORGANIZATION IN HIGHER VERTEBRATES

Before discussing the ways of blood proteins organization in fishes, let us consider some general conceptions about extracellular fluids of the organism, blood plasma proteins localization in the organism, principles of organization of extracellular (blood plasma proteins) and intracellular (hemoglobin) blood proteins in mammals; consider the structural models of some proteins, with particular attention to the way of organization of proteins by the type of monomer/oligomer and the role of formation and degradation of oligomeric proteins in osmoregulation.

1.1. IN WHAT EXTRACELLULAR FLUIDS OF THE ORGANISM DO BLOOD PLASMA PROTEINS OCCUR?

Organism Internal Fluid Environment

The internal environment of the organism includes interstitial fluid, blood plasma, lymph and some other extracellulur fluids, separated by vascular walls. Its role is to ensure the homeostasis of all organism physiological functions. In addition during the early development the internal fluid environment of the organism actively affects the embryo gene expression (Figure 1.1) (Andreeva, 2005a, b, 2007, 2009, 2011).

The interstitial fluid washes organism cells, blood plasma integrates all extracellular fluids and the lymph returns filtrates of the tissue fluids into the bloodstream. In vertebrates with closed circulatory system the blood plasma proteins are localized mainly inside the circulatory system vessels. The exchange of both fluids and proteins included in them is going through the capillaries' walls. Proteins penetrate the interstitial space through the capillaries' walls and then return into the blood bed through the drainage lymphatic system.

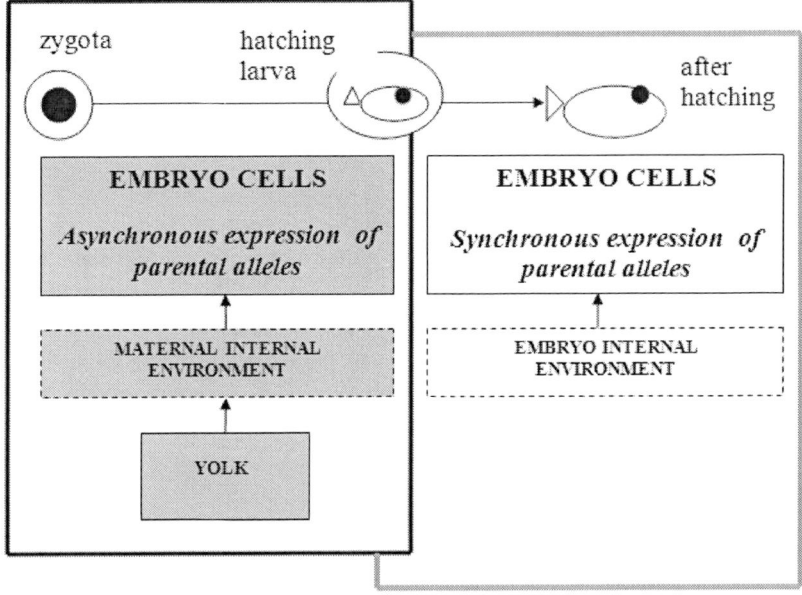

Figure 1.1. Influence of yolk to character of expression of parental alleles of embryo genes from intergeneric F1 hybrids of bream, roach and zope. Compartments marked by grey color contain maternal yolk proteins (Andreeva, 2009, 2011).

What Proteins Are Included in the Blood Plasma, Serum and Interstitial Fluid of the Organism?

Blood plasma includes the following proteins: albumins, globulins and fibrinogen. The latter is a part of the blood coagulation system. Blood serum differs from plasma by the absence of fibrinogen. Some mammalian interstitial fluids also contain blood plasma proteins.

Albumins and globulins differ in their solubility in saline solutions. So, in ammonium sulphate semisaturated solution albumins remain in the solubilized state, unlike less stable to dehydratationglobulins, which precipitate. The following proteins are globulins: $α_2$-globulin haptoglobin, β-globulins transferrin and hemopexin, γ-globulins immunoglobulins.

Albumin is osmotically active, transport and plastic protein. Globulins functions are very diverse: transferrin is the Fe^{3+}- ion carrier, haptoglobin binds extracellular hemoglobin, hemopexin binds hemin, and immunoglobulins, unlike all other proteins, bind the maximum possible number of ligands.

1.2. THE ROLE OF BLOOD PLASMA PROTEINS IN WATER METABOLISM OF THE ORGANISM

The processes of interstitial fluid filtration in the organism are determined mainly by the blood plasma proteins. The main function of plasma proteins appears to be water retention, therefore, the proteins spread between intra- and extravascular fluids of the organism so that the balance of extracellular fluid all over the organism is achieved and comfortable osmotic conditions for its all cells functioning are maintained. Albumin is osmotically active blood protein. Its small sizes and high solubility make possible its filtration through capillary walls into the interstitial space and to perform the transport function along with the osmoregulation.

Classical conceptions about capillary fluid exchange nature in animals with closed blood circulation are based on the Starling hypothesis (Starling, 1895). According to this hypothesis extracellular fluid balance in the organism is maintained by intravascular fluid filtration through the capillary wall into the interstitial space under the influence of hydrostatic pressure (Figure 1.2). This pressure is counteracted by the oncotic pressure of blood plasma, created by proteins, which pump the water from the extracellular space (Figure 1.3). If inside the blood vessels there were no proteins, then any hydrostatic pressure, which exceeds the pressure of tissue fluid, would force fluid to be filtered outside and there would be no force, which makes fluid to return conversely. Namely proteins support the volume of intravascular fluid at the specific level (Figure 1.4). Despite the fact that the amount of fluid, which emerges through the walls of capillaries, and the amount of fluid, entering conversely, strongly varies, within the standard the output flow exceeds the input one. In the organism transvascular fluid flows are directed from the vessels toward the tissues (Papenfuss, Hauck, 1987).

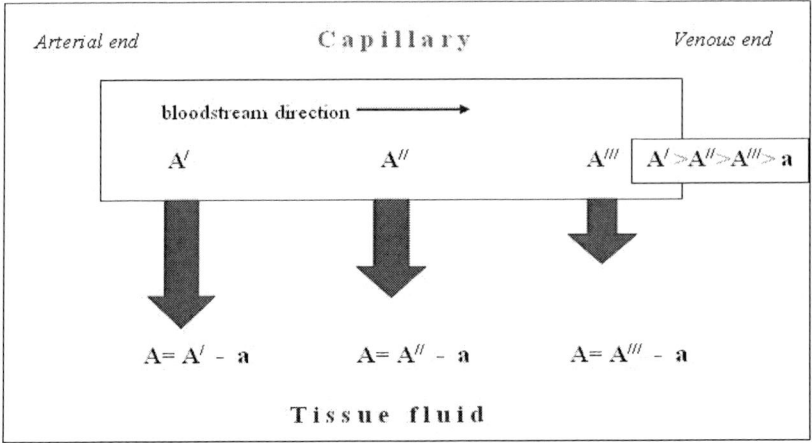

Figure 1.2. Hydrostatic pressure within the capillary ($A^{/}$, $A^{//}$, $A^{///}$) and in tissue fluid (a).

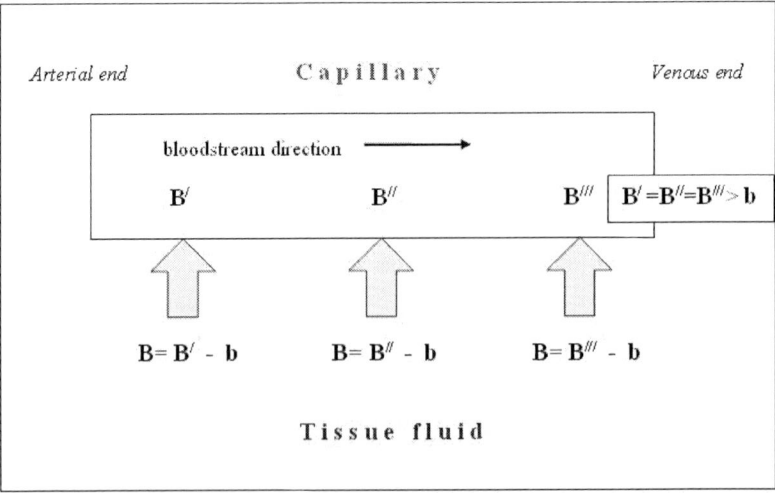

Figure 1.3. Oncotic pressure of plasma ($B^{/}$, $B^{//}$, $B^{///}$) and tissue fluid (b).

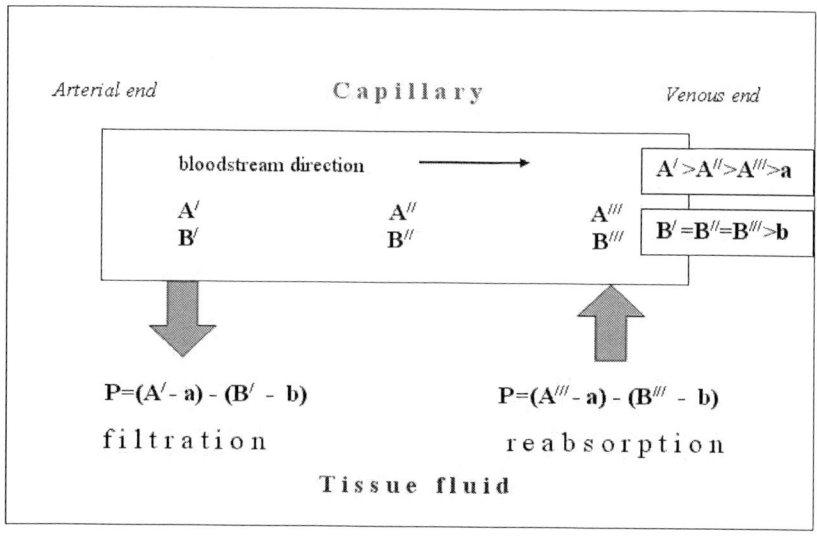

Figure 1.4. The result of interaction of hydrostatic and oncotic pressures into capillary (A, B) and in tissue fluid (a, b).

The Starling conception is based on the assumption that blood plasma proteins are localized only inside the blood vessels and aren't able to penetrate into interstitial space. However, it was defined that blood plasma proteins penetrate into the interstitial space by means of diffusion through the pores between endothelial cells and through the endothelial cells directly (Papenfuss, Gross, 1985, 1987; Papenfuss, Hauck, 1987; Adamson et al., 2004; Ostensson, Lun, 2008; Sarin, 2010). It is determined for the albumin that its transvascular transport is provided by the convection through so-called big pores in 75-90% cases, the rest of the albumin got into the interstitial fluid by means of diffusion through the small pores; larger proteins got from the blood into the interstitial fluid by means of convection through the big pores exclusively (Rippe, Haraldsson, 1987). Transendothelial transport of proteins is realized by means of membrane-bound tubular-vesicular system and is considered as a counterpart of the transport through the big pores (Bendayan, Rasio, 1997). Endothelial cells can play an active role in the transvascular protein exchange by changing their phenotype in response to physical and chemical stress (Curry, 2005). Reversal transfer of the proteins (albumin) from the interstitial fluid into the vessels is prevented by ultrafiltrate flowing from the intravascular space through the pores between endothelial cells (Adamson et al., 2004).

Mammalian blood plasma proteins are filtered into the interstitial space not in all the parts of the capillary network: so, the brain capillaries are appeared to be completely tight impermeable, and the liver capillaries – completely permeable for blood plasma proteins; muscle type capillaries have intermediate permeability values for blood plasma proteins (Gamble, 1954; Landis, Pappenheimer, 1963; Scholander et al., 1968; Zweifach, Intaglietta, 1968; Guyton et al., 1971; Polenov, Dvoreckij, 1976; White et al., 1978).

1.3. WHAT SHOULD BLOOD PLASMA PROTEINS BE LIKE IN ANIMALS WITH CLOSED BLOOD CIRCULATORY SYSTEM?

Size of the Blood Plasma Proteins

Intravascular localization and transcapillary exchange restrict the size of blood plasma proteins: the sizes should be such that the proteins were trapped inside the blood vessels and did not filter easily through the capillary wall. This characteristic is typical for proteins with molecular weights not less than 60 kDa possess (Shulz, Schirmer, 1979). Meanwhile, blood proteins shouldn't be too large in order not to increase blood viscosity. As for surface structure of blood plasma proteins, it should allow them to carry metabolites and perform other specific functions.

Monomeric and Oligomeric Proteins

In addition to the molecular weights that is above 60 kDa true blood proteins should be monomerous, i.e. to consist of a single polypeptide chain or several chains, linked one to another with covalent bonds. If the protein is oligomeric, i.e. consists of several polypeptide chains linked with a noncovalent bond, then after it dissociates into subunits, the latter will filtrate through the capillary walls easily, resulting in blood oncotic pressure reduction and, therefore, in interstitial fluid balance disturbance. Some proteins increase their size not by adding amino acids but due to binding the carbohydrates covalently to them.

1.4. STRUCTURAL MODELS OF BLOOD PLASMA PROTEINS AND HEMOGLOBIN IN MAMMALS

Serum Albumin

Albumin belongs to the protein superfamily which also includes alpha-fetoprotein, vitamin D-binding protein and afamin (family:"ALB/AFP/VDB family" in UniProtKB; Schoentgen et al., 1986; Haefliger et. al., 1989; Lichenstein et. al., 1994). Human seralbumin gene is located in the 4th chromosome (NCBI Reference Sequence: ACCESSION NG_009291) and is presented by a single copy (Hawkins, Dugaiczyk, 1982; Minghetti et al., 1986). Human serum albumin (HSA) consists of 585 amino acids and is formed from the precursor preproalbumin, which consists of 609 amino acids (Dugaiczyk et. al., 1982; Minchiotti et. al., 2008; UniProtKB/Swiss-Prot: P02768 (ALBU_HUMAN)).

The 3D structure of HSA has been determined by X-ray crystallography to a resolution of 2.8A (PUBMED:1630489). The albumin molecule consists of three domains and the spiral structure. Each domain is a product of two subdomains that possess common structural motifs (PUBMED:1630489). Two subunits are connected by S-S-bonds (Klotz et al., 1975; Kragh-Hansen, 1990; He, Carter, 1992). These three domains are homologous, they form a heart-shaped molecule (PUBMED:1630489) (Figure 1.5), which varies under different factors (Pantjavin et al., 2000).

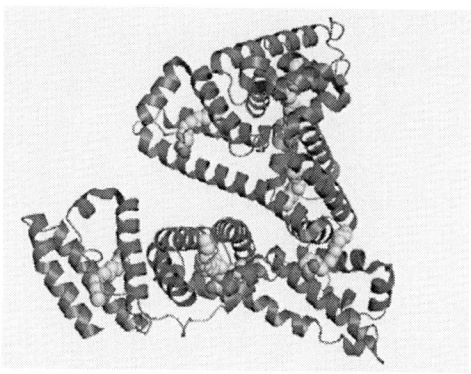

Figure 1.5. The model of structure of Human serum albumin (HSA) complexed with 6 palmitic acid molecules (PDB).

Each domain of HSA contains five or six internal disulphide bonds (Figure 1.6).

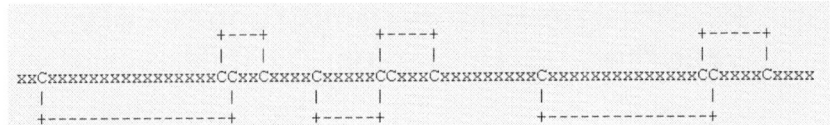

Figure 1.6. The scheme of location of S-S-bonds in domain of human serum albumin (EMBL-EBI: IPR014760 Serum albumin, N-terminal).

Altogether in the structure of albumin there are 17 S-S-bonds, which stabilize the single polypeptide chain (Saber et al., 1977) (Figure 1.7), and one free SH-group (Sugio et al., 1999).

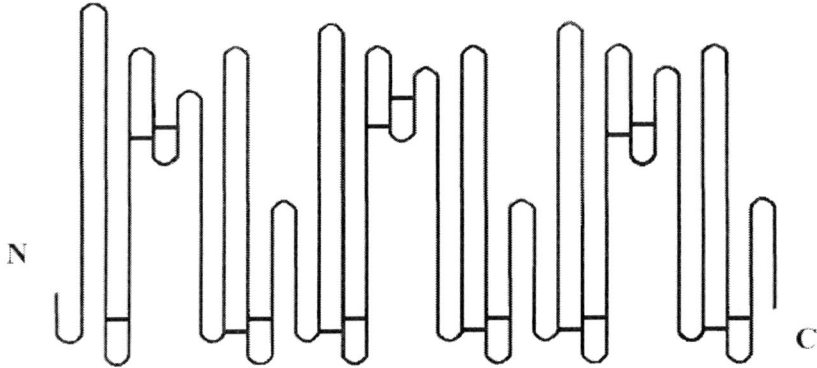

Figure 1.7. The scheme of location of S-S-bonds in molecule of human serum albumin (according to Behrens et al., 1974).

Thecarbohydrates are absent in the structure of this protein, with the exception of pathological cases – diabetes disease (UniProtKB/Swiss-Prot: P02768 (ALBU_HUMAN)), in mutant albumin variants also, among them albumin Kenitra (Minchiotti et al, 2001; Kragh-Hansen et al., 2005), Casebrook (Peach, Brennan, 1991; Nielsen et al., 1997), Dalakarlia-1 (Carlson et al., 1992; Arai et al., 1990), Redhill (Brand et al 1984; Hutchinson and Matejtschuk, 1985; Nielsen et al., 1997; Brennan et al., 1990; Minchiotti et al, 2008). Thus, the allotypes of three mutant albumins undergo to N-glycosylation, one C- end mutation of albumin with the shift of the reading frame led to O- glycosylation. The elongate albumin Kenitra relates to glycated mutant versions of albumin (Minchiotti et al, 2008).

Being monomeric, the albumin can form *in vivo* dimers, trimers and tetramers with MM about 130, 195 and 260 kDa, respectively, by means of free surface SH-group. The formed structures are also monomeric proteins. In SDS-PAGE (reducing conditions) such proteins should dissociate into monomers, however, there is evidence that not all of the dimmers dissociate into monomers. The reason for this is presumably concealed in heterogeneity of albumins, one of manifestations of which is nonidentity of free SH-groups in different protein molecules. The glycosilation of albumin leads to formation of high-molecular aggregates (Ali Han et al., 2007).

Mammalian albumins participate in the transport of hormones, fat acids, palmitic acid, various metabolites; Ca^{2+}, Na^+, K^+, Zn^{2+} (UniProtKB/Swiss-Prot: P02768 (ALBU_HUMAN); Lu et. al., 2008), and, unlike fishes, - nickel (Minchiotti et al, 2008). HSA have a high affinity to nickel Ni (II)andcuprum Cu(II). The binding occur with participation of N- terminal sequence Asp-Ala-His. The disorder of this site leads to low binding of these metalions (Minchiotti et al., 2008). The binding of metals by albumins leads to change of 3D structure of protein.

In the electrophoresis mammal albumins are presented, as a rule, by one macrocomponent, since the level of real homozygocity in the locus of albumin reaches 90% in mammalian(Tinaeva et al., 2007). However, the plural versions of protein are manifested in heterozygotes at some mutations, because of the codominant expression of albumin gene alleles. 77 mutations of albumin are known in human, the most of them are generally harmless. In the majority of the cases they consist in the replacement of the charged amino-acid residues and are revealed in the electrophoresis. 65 mutations of 77 known (Minchiotti et al., 2008) are manifested as bisalbuminemia in electrophoresis in the form of additional bands at the electrophoregram; 12 mutations - Analb Baghdad (Campagnoli et al., 2002), Analb Codogno (Watkins et al., 1994b), Analb Kauseri (Galliano et al., 2002) and others (Minchiotti et al., 2008) - lead to decrease in the content of albumin in the blood or analbuminemia (Minchiotti et al, 2008).

Albumins from other mammals possess resembling structures and properties: bovine *Bos taurus* (Patterson, Geller, 1977; Hirayama et al., 1990; UniProtKB/Swiss-Prot: P02769 (ALBU_BOVIN)), rat *Rattus norvegicus* (Strauss et al., 1997; Sargent et al., 1981; UniProtKB/Swiss-Prot: P02770 (ALBU_RAT)), mouse *Mus musculus* (UniProtKB/Swiss-Prot: P07724 (ALBU_MOUSE)), horse *Equus caballus* (Lambert, Kelly, 1978; Ho et al., 1993), sheep*Ovis aries* (Brown et al., 1989; Thompson et al., 1992) (Figure 1.8).

Sequence Alignment of the Serum Albumin Protein (with Clustalw)

Species Sequence Alignment of Serum Albumin

```
Mouse    MKWVTFLLLLFVSGSAFSR---GVFRREA---HKSEIAHRYNDLGEQHFKGLVLIAFSQY 54
Rat      MKWVTFLLLLFISGSAFSR---GVFRREA---HKSEIAHRFKDLGEQHFKGLVLIAFSQY 54
Dog      MKWVTFISLFFLFSSAYSR---GLVRREA---YKSEIAHRYNDLGEEHFRGLVLVAFSQY 54
Cat      MKWVTFISLLLLFSSAYSR---GVTRREA---HQSEIAHRFNDLGEEHFRGLVLVAFSQY 54
Human    MKWVTFISLLFLFSSAYSR---GVFRRDA---HKSEVAHRFKDLGEENFKALVLIAFAQY 54
Cow      MKWVTFISLLLLFSSAYSR---GVFRRDT---HKSEIAHRFKDLGEEHFKGLVLIAFSQY 54
Sheep    MKWVTFISLLLLFSSAYSR    GVFRRDT   HKSEIAHRFNDLGEENFQCLVLIAFSQY 54
Pig      MKWVTFISLLFLFSSAYSR---GVFRRDT---YKSEIAHRFKDLGEQYFKGLVLIAFSQH 54
Horse    MKWVTFVSLLFLFSSAYSR---GVLRRDT---HKSEIAHRFNDLGEKHFKGLVLVAFSQY 54
Rabbit   MKWVTFISLLFLFSSAYSR---GVFRREA---HKSEIAHRFNDVGEEHFIGLVLITFSQY 54
Chicken  MKWVTLISFIFLFSSATSRNLQRFARDAE---HKSEIAHRYNDLKEETFKAVAMITFAQY 57
Frog     MKWITLICLLISSTLIESR---IIFKRDTDVDHHKHIADMYNLLTERTFKGLTLAIVSQN 57
                           **     . :       ::..:*.  :: :  *. * .:.:   .:*

Mouse    LQKCSYDEHAKLVQEVTDFAKTCVADESAANCDKSLHTLFGDKLCAIPNLRENYGELADC 114
Rat      LQKCPYEEHIKLVQEVTDFAKTCVADENAENCDKSIHTLFGDKLCAIPKLRDNYGELADC 114
Dog      LQQCPFEDHVKLAKEVTEFAKACAAEESGANCDKSLHTLFGDKLCTVASLRDKYGDMADC 114
Cat      LQQCPFEDHVKLVNEVTEFAKGCVADQSAANCEKSLHELLGDKLCTVASLRDKYGEMADC 114
Human    LQQCPFEDHVKLVNEVTEFAKTCVADESAENCDKSLHTLFGDKLCTVATLRETYGEMADC 114
Cow      LQQCPFDEHVKLVNELTEFAKTCVADESHAGCEKSLHTLFGDELCKVASLRETYGDMADC 114
Sheep    LQQCPFDEHVKLVKELTEFAKTCVADESHAGCDKSLHTLFGDELCKVATLRETYGDMADC 114
Pig      LQQCPYEEHVKLVPEVTEFAKTCVADESAENCDKSIHTLFGDKLCAIPSLREHYGDLADC 114
Horse    LQQCPFEDHVKLVNEVTEFAKKCAADESAENCDKSLHTLFGDKLCTVATLRATYGELADC 114
Rabbit   LQKCPYEEHAKLVKEVTDLAKACVADESAANCDKSLHDIFGDKICALPSLRDTYGDVADC 114
Chicken  LQRCSYEGLSKLVKDVVDLAQKCVANEDAPECSKPLPSIILDEICQVEKLPDSYGAMADC 117
Frog     LQKCSLEELSKLVNEINDFAKSCTGNDKTPECEKPIGTLFYDKLCADPKVGVNYEWSKEC 117
          **:*. :  **..::  ::*: *..::.   *.*.:   :: *::*    .:    *    :*

Mouse    CTKQEPERNECFLQHKDDNPSLP-PFERPEAEAMCTSFKENPTTFMGHYLHEVARRHPYF 173
Rat      CAKQEPERNECFLQHKDDNPNLP-PFQRPEAEAMCTSFQENPTSFLGHYLHEVARRHPYF 173
Dog      CEKQEPDRNECFLAHKDDNPGFP-PLVAPEPDALCAAFQDNEQLFLGKYLYEIARRHPYF 173
Cat      CEKKEPERNECFLQHKDDNPGFG-QLVTPEADAMCTAFHENEQRFLGKYLYEIARRHPYF 173
Human    CAKQEPERNECFLQHKDDNPNLP-RLVRPEVDVMCTAFHDNEETFLKKYLYEIARRHPYF 173
Cow      CEKQEPERNECFLSHKDDSPDLP-KLK-PDPNTLCDEFKADEKKFWGKYLYEIARRHPYF 172
Sheep    CEKQEPERNECFLNHKDDSPDLP-KLK-PEPDTLCAEFKADEKKFWGKYLYEVARRHPYF 172
Pig      CEKKEPERNECFLQHKNDNPDIP-KLK-PDPVALCADFQEDEQKFWGKYLYEIARRHPYF 172
Horse    CEKQEPERNECFLTHKDDHPNLP-KLK-PEPDAQCAAFQEDPDKFLGKYLYEVARRHPYF 172
Rabbit   CEKKEPERNECFLHHKDDKPDLP-PFARPEADVLCKAFHDDEKAFFGHYLYEVARRHPYF 173
Chicken  CSKADPERNECFLSFKVSQPDFVQPYQPPASDVICQEYQDNRVSFLGHFIYSVARRHPFL 177
Frog     CSKQDPERAQCFRAHFVFEHNPV----RPKPEETCALFKEHPDDLLSAFIHEEAPNHPDL 173
          * *  :*:* :**      .:  .         *    *  ::  .   :   :::.  **.**  :

Mouse    YAPELLYYAEQYNEILTQCCAEADKESCLTPKLDGVKEKALVSSVRQRMKCSSMQKFGER 233
Rat      YAPELLYYAEKYNEVLTQCCTESDKAACLTPKLDAVKEKALVAAVRQRMKCSSMQRFGER 233
Dog      YAPELLYYAQQYKGVFAECCQAADKAACLGPKIEALREKVLLSSAKERFKCASLQKFGDR 233
Cat      YAPELLYYAEEYKGVFTECCEAADKAACLTPKVDALREKVLASSAKERLKCASLQKFGER 233
Human    YAPELLFFAKRYKAAFTECCQAADKAACLLPKIDELRDEGKASSAKQRLKCASLQKFGER 233
Cow      YAPELLYYANKYNGVFQECCQAEDKGACLLPKIETMREKVLTSSARQPLRCASIQKFGER 232
Sheep    YAPELLYYANKYNGVFQECCQAEDKGACLLPKIDAMREKVLASSARQRLRCASIQKFGER 232
Pig      YAPELLYYAIIYKLVFSECCQAADKAACLLPKIEHLREKVLTSAAKQRLKCASIQKFGER 232
Horse    YGPELLFHAEEYKADFTECCPADDKLACLIPKLDALKERILLSSAKEPLKCSSFQNFGER 232
Rabbit   YAPELLYYAQKYKAILTECCEAADKGACLTPKLDALEGKSLISAAQERLRCASIQKFGDR 233
Chicken  YAPAILSFAVDFEHALQSCCKESDVGACLDTKEIVMREKAKGVSVKQQYFCGILKQFGDR 237
Frog     YPPAVLLLTQQYGKLVEHCCEEEDKDKCFAEKMKELMKHSHSIEDKQKHFCWIVNNYPER 233
         * *  :*  :   :      .   **   *  *:*   * :   .       :::  *  .:.:  :*
```

Principles of the Blood Plasma Proteins Organization ... 13

```
Mouse    AFKAWAVARLSQTFPNADFAEITKLATDLTKVNKECCHGDLLECADDRAELAKYMCENQA 293
Rat      AFKAWAVARMSQRFPNAEFAEITKLATDLTKINKECCHGDLLECADDRAELAKYMCENQA 293
Dog      AFKAWSVARLSQRFPKADFAEISKVVTDLTKVHKECCHGDLLECADDRADLAKYMCENQD 293
Cat      AFKAWSVARLSQKFPKAEFAEISKLVTDLAKIHKECCHGDLLECADDRADLAKYICENQD 293
Human    AFKAWAVARLSQRFPKAEFAEVSKLVTDLTKVHTECCHGDLLECADDRADLAKYICENQD 293
Cow      ALKAWSVARLSQKFPKAEFVEVTKLVTDLTKVHKECCHGDLLECADDRADLAKYICDNQD 292
Sheep    ALKAWSVARLSQKFPKADFTDVTKIVTDLTKVHKECCHGDLLECADDRADLAKYICDHQD 292
Pig      AFKAWSLARLSQRFPKADFTEISKIVTDLAKVHKECCHCDLLECADDRADLAKYICENQD 292
Horse    AVKAWSVARLSQKFPKADFAEVSKIVTDLTKVHKECCHGDLLECADDRADLAKYICEHQD 292
Rabbit   AYKAWALVRLSQRFPKADFTDISKIVTDLTKVHKECCHGDLLECADDRADLAKYMCEHQE 293
Chicken  VFQARQLIYLSQKYPKAPFSEVSKFVHDSIGVHKECCEGDMVECMDDMAPMMSNLCSQQD 297
Frog     VIKALNLARVSHRYPKPDFKLAHKFTEETTHFIKDCCHGDMFECMTERLELSEHTCQHKD 293
         . :*   :  :*: :*:.  *      *..  :  . ..:**.**:.**  :  : .  *.::

Mouse    TISSKLQTCCDKPLLKKAHCLSEVEHDTMPADLPAIAADFVEDQEVCKNYAEAKDVFLGT 353
Rat      TISSKLQACCDKPVLQKSQCLAEIEHDNIPADLPSIAADFVEDKEVCKNYAEAKDVFLGT 353
Dog      SISTKLKECCDKPVLEKSQCLAEVERDELPGDLPSLAADFVEDKEVCKNYQEAKDVFLGT 353
Cat      SISTKLKECCGKPVLEKSHCISEVERDELPADLPPLAVDFVEDKEVCKNYQEAKDVFLGT 353
Human    SISSKLKECCEKPLLEKSHCIAEVENDEMPADLPSLAADFVESKDVCKNYAEAKDVFLGM 353
Cow      TISSKLKECCDKPLLEKSHCIAEVEKDAIPENLPPLTADFAEDKDVCKNYQEAKDAFLGS 352
Sheep    ALSSKLKECCDKPVLEKSHCIAEVDKDAVPENLPPLTADFAEDKEVCKNYQEAKDVFLGS 352
Pig      TISTKLKECCDKPLLEKSHCIAEAKRDELPADLNPLEHDFVEDKEVCKNYKEAKHVFLGT 352
Horse    SISGKLKACCDKPLLQKSHCIAEVKEDDLPSDLPALAADFAEDKEICKHYKDAKDVFLGT 352
Rabbit   TISSHLKECCDKPILEKAHCIYGLHNDETPAGLPAVAEEFVEDKDVCKNYEEAKDLFLGK 353
Chicken  VFSGKIKDCCEKPIVERSQCIMEAEFDEKPADLPSLVEKYIEDKEVCKSFEAGHDAFMAE 357
Frog     ELSTKLEKCCNLPLLERTYCIVTLENDDVPAELSKPITEFTEDPHVCEKYAENK-SFLEI 352
         :*  ::: **  *::::: *:   . *   *       .: *. ..:*:  :   *:

Mouse    FLYEYSRRHPDYSVSLLLRLAKKYEATLEKCCAEANPPACYGTVLAEFQPLVEEPKNLVK 413
Rat      FLYEYSRRHPDYSVSLLLRLAKKYEATLEKCCAEGDPPACYGTVLAEFQPLVEEPKNLVK 413
Dog      FLYEYAFRHPEYSVSLLLRLAKEYEATLEKCCATDDPPTCYAKVLDEFKPLVDEPQNLVK 413
Cat      FLYEYSRRHPEYSVSLLLRLAKEYEATLEKCCATDDPPACYAHVFDEFKPLVEEPHNLVK 413
Human    FLYEYARRHPDYSVVLLLRLAKTYETTLEKCCAAADPHECYAKVFDEFKPLVEEPQNLIK 413
Cow      FLYEYSRRHPEYAVSVLLRLAKEYEATLEECCAKDDPHACYSTVFDKLKHLVDEPQNLIK 412
Sheep    FLYEYSRRHPEYAVSVLLRLAKEYEATLEDCCAKEDPHACYATVFDKLKHLVDEPQNLIK 412
Pig      FLYEYSRRHPDYSVSLLLRIAKIYEATLEDCCAKEDPPACYATVFDKFQPLVDEPKNLIK 412
Horse    FLYEYSRRHPDYSVSLLLRIAKTYEATLEKCCAEADPPACYRTVFDQFTPLVEEPKSLVK 412
Rabbit   FLYEYSRRHPDYSVVLLLRLGKAYEATLKKCCATDDPHACYAKVLDEFQPLVDEPKNLVK 413
Chicken  FVYEYSRRHPEFSIQLIMRIAKGYESLLEKCCKTDNPAECYANAQEQLNQHIKETQDVVK 417
Frog     SPWQ-SQETPELSEQFLLQSAKEYESLLNKCCFSDNPECYKDGADRFMNEAKERFAYLK 411
         ::  ::. *:  :  .::: .* **: *:.**   :* **     .:    .*    :*

Mouse    TNCDLYEKLGEYGFQNAILVRYTQKAPQVSTPTLVEAARNLGRVGTKCCTLPEDQRLPCV 473
Rat      TNCELYEKLGEYGFQNAILVRYTQKAPQVSTPTLVEAARNLGRVGTKCCTLPEAQRLPCV 473
Dog      TNCELFEKLGEYGFQNALLVRYTKKAPQVSTPTLVEVSRKLGKVGTKCCKKPESERMSCA 473
Cat      TNCELFEKLGEYGFQNALLVRYTKKVPQVSTPTLVEVSRSLGKVGSKCCTHPEAERLSCA 473
Human    QNCELFEQLGEYKFQNALLVRYTKKVPQVSTPTLVEVSRNLGKVGSKCCKHPEAKRMPCA 473
Cow      QNCDQFEKLGEYGFQNALIVRYTRKVPQVSTPTLVEVSRSLGKVGTRCCCTKPESERMPCT 472
Sheep    KNCELFEKHGEYGFQNALIVRYTRKAPQVSTPTLVEISRSLGKVGTKCCAKPESERMPCT 472
Pig      QNCELFEKLGEYGFQNALIVRYTKKVPQVSTPTLVEVARKLGLVGSRCCKRPEEERLSCA 472
Horse    KNCDLFEEVGEYDFQNALIVRYTKKAPQVSTPTLVEIGRTLGKVGSRCCKLPESERLPCS 472
Rabbit   QNCELYEQLGDYNFQNALLVRYTKKVPQVSTPTLVEISRSLGKVGSKCCKHPEAERLPCV 473
Chicken  TNCDLLHDHGEADFLKSILIRYTKKMPQVPTDLLLETGKKMTTIGTKCCQLGEDRPMACS 477
Frog     QNCDILHEHGEYLFENELLIRYTKKMPQVSDETLIGIAHQMADIGEHCCAVPENQRMPCA 471
         **:  .. *: * : :::***:* ***. *: .: :  :* :**   * .*:.*

Mouse    EDYLSAILNRVCLLHEKTPVSEHVTKCCSGSLVERRPCFSALTVDETYVPKEFKAETFTF 533
Rat      EDYLSAILNRLCVLHEKTPVSEKVTKCCSGSLVERRPCFSALTVDETYVPKEFKAETFTF 533
Dog      EDFLSVVLNRLCVLHEKTPVSERVTKCCSESLVNRRPCFSGLEVDETYVPKEFNAETFTF 533
```

```
Cat     EDYLSVVLNRLCVLHEKTPVSERVTKCCTESLVNRRPCFSALQVDETYVPKEFSAEIFTF 533
Human   EDYLSVVLNQLCVLHEKTPVSDRVTKCCTESLVNRRPCFSALEVDETYVPKEFNAETFTF 533
Cow     EDYLSLILNRLCVLHEKTPVSEKVTKCCTESLVNRRPCFSALTPDETYVPKAFDEKLFTF 532
Sheep   EDYLSLILNRLCVLHEKTPVSEKVTKCCTESLVNRRPCFSDLTLDETYVPKPFDEKFFTF 532
Pig     EDYLSLVLNRLCVLHEKTPVSEKVTKCCTESLVNRRPCFSALTPDETYKPKEFVEGIFTF 532
Horse   ENHLALALNRLCVLHEKTPVSEKITKCCTDSLAERRPCFSALELDEGYVPKEFKAETFTF 532
Rabbit  EDYLSVVLNRLCVLHEKTPVSEKVTKCCSESLVDRRPCFSALGPDETYVPKEFNAEIFTF 533
Chicken EGYLSIVIHDTCFKQETTPINDNVSQCCSQLYANRRPCFTAMGVDTKYVPPPFNPDMFSF 537
Frog    EGDLTILIGKMCERQKKTFINNHVAHCCTDSYSGMRSCFTALGPDEDYVPPPVTDDIFHF 531
        *. *:   :   *   ::.*  :.:.:::**:          *.**:  :  *  * *  .     * *

Mouse   HSDICTLPEKEKQIKKQTALAELVKHKPKATAEQLKTVMDDFAQFLDTCCKAADKDICFS 593
Rat     HSDICTLPDKEKQIKKQTALAELVKHKPKATEDQLKTVMGDFAQFVDKCCKAADKDNCFA 593
Dog     HADLCTLPEAEKQVKKQTALVELLKHKPKATDEQLKTVMGDFGAFVEKCCAAENKEGCFS 593
Cat     HADLCTLPEAEKQIKKQSALVELLKHKPKATEEQLKTVMGDFGSFVDKCCAAEDKEACFA 593
Human   HADICTLSEKERQIKKQTALVELVKHKPKATKEQLKAVMDDFAAFVEKCCKADDKETCFA 593
Cow     HADICTLPLTEKQIKKQTALVELLKHKPKATEEQLKTVMENFVAFVDKCCAADDKEACFA 592
Sheep   HADICTLPDTEKQIKKQTALVELLKHKPKATDEQLKTVMENFVAFVDKCCAADDKEGCFV 592
Pig     HADLCTLPEDEKQIKKQTALVELLKHKPHATEEQLRTVLGNFAAFVQKCCAAPDHEACFA 592
Horse   HADICTLPEDEKQIKKQSALAELVKHKPKATKEQLKTVLGNFSAFVAKCCGREDKEACFA 592
Rabbit  HADICTLPETERKIKKQTALVELVKHKPHATNDQLKTVVGEFTALLDKCCSAEDKEACFA 593
Chicken DEKLCSAPAEEREVGQMKLLINLIKRKPQMTEEQIKTIADGFTAMVDKCCKQSDINTCFG 597
Frog    DDKICTANIKEKQHIKQKFLVKLIKVSPKLEKNHIDEWLLEFLKMVQKCCTADEHQPCFD 591
        . .:*:      *::   :.  * :*:*  .*:    :::         *  ::  .**    : : **

Mouse   TEGPNLVTRCKDTLA--- 608
Rat     TEGPNLVARSKEALA--- 608
Dog     EEGPKLVAAAQAALV--- 608
Cat     EEGPKLVAAAQAALA--- 608
Human   EEGKKLVAASQAALGL-- 609
Cow     VEGPKLVVSTQTALA--- 607
Sheep   LEGPKLVASTQAALA--- 607
Pig     VEGPKFVIEIRGILA--- 607
Horse   EEGPKLVASSQLALA--- 607
Rabbit  VEGPKLVESSKATLG--- 608
Chicken EEGANLIVQSRATLGIGA 615
Frog    TEKPVLIEHCQKLHP--- 606
              *    ::   :
```

Accession numbers of the serum albumin sequences used in the protein sequence alignment :NP_000468 - Human; NP_033784 - Mouse; NP_599153 - Rat; NP_851335 - Cow; NP_001005208 - Pig; NP_001003026 - Dog; NP_001009961 - Cat; NP_001009376 - Sheep; NP_001075972 - Horse; NP_001075813 - Rabbit; NP_990592 - Chicken; NP_001081244 – Frog.

Figure 1.8. Amino acid sequences of serum albumins of some vertebrata (Higgins et al., 1994).

Mammal albumins can be briefly characterized as simple monomeric proteins with low molecular weights capable of forming more large monomers by means of surface SH-group; under the pathology conditions the serum albumin concentration in the blood can substantially decrease; and glycated proteins, plural and oligomeric albumin appear among structural variants of albumins.

Immunoglobulins

Immunoglobulins belong to the superfamily of proteins with Y-shaped spatial organization, capable to bind antigens, which promote their generation. Immunoglobulins are produced by B-lymphocytes and are situated either as free forms in blood or as receptors on cell surface membranes. All the immunoglobulins consist of two heavy (H) chains and two light (L) chains, connected by S-S-bonds: each H-chain is connected to L-chain by one S-S-bond, and two H-chains are connected by two S-S-bonds (Figure 1.9).

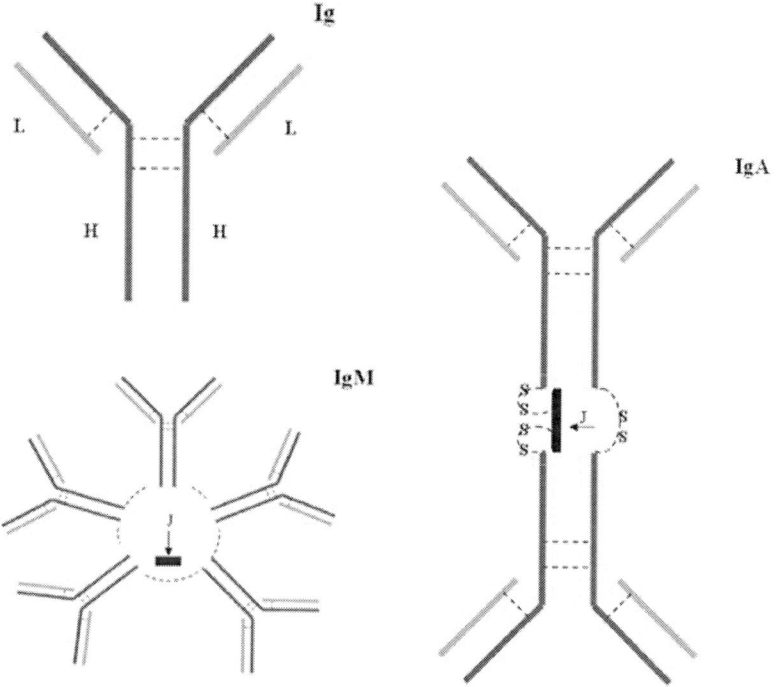

Figure 1.9. Structure of immunoglobulin molecule consist of one unit and several units.

Five immunoglobulin types are known in mammals: G, A, M, D and E (Table 1.1).

Table 1.1. Immunoglobulin types and the chains (H and L) in their composition

Ig and its chains	Ig type				
	IgG	IgA	IgM	IgD	IgE
H-chain MM	γ 53 000	α 64 000	μ 70 000	δ 58 000	ε 75 000
L-chain MM	κ or λ 22 500				
J-chain MM	-	15 000	15 000	-	-
structural formula of Ig	H_2L_2	$(H_2L_2)nJ$ n=2, 4	$(H_2L_2)nJ$ n=5 or multiply of 5	H_2L_2	H_2L_2
MM	150 000	360 000 720 000	935 000	172 000	196 000

Different types of Ig vary in sedimentation and electrophoretic characteristics due to different H-chain composition. Mammalian immunoglobulins are monomeric proteins and have carbohydrates in their structure. IgA and IgM consist of several monomeric units, held together by S-S-bonds by means of J-chain; IgG, IgE and IgD each have one monomeric unit (Figure 1.9) (Rogers et al., 2006).

The genes encoding light and heavy chains assemble from separate segments during the lymphocytes' maturation. Mammals have nearly 300 segments in their genome, encoding variable domains of immunoglobulin light chains (V_L) (first 95 amino acids), several (up to 6) J_L-segments (amino acids 96-107) and C_L-segments, encoding the constant domains of L-chains (amino acids 108-214). Heavy chain genes also consist of variable segments (V_H) and of D (10-50), J (4) and C-segments (8); the number of C-segments correspond to the number of immunoglobulin heavy chain types. The genes (segments), encoding human immunoglobulin, are localized in 2nd, 14th and 22nd chromosomes (NCBI GenBank: U79587.1; NCBI Reference Sequence: NM_152855.1).

In the early stages of evolution the genes of immunoglobulins were organized in the form of repetitive clusters V-J-C of genetic segments. In this organization the variety of the produced antibodies was based, first of all, on a quantity of clusters available in the genome, each of which encoded one variant of the polypeptide subunits of immunoglobulins. In the process of further evolution the transition from the cluster to the segmental organization occurrs, which ensures the additional source of variety due to the combinatorial recombination of genetic segments.

Haptoglobin

α_2-globulins haptoglobins belong to the S1 peptidase family (family: "peptidase S1 family" in UniProtKB). Haptoglobins form specific complexes *in vivo* with hemoglobin, which enters the blood stream in result of intravascular hemolysis of erythrocytes.

Human haptoglobins consist of two pair of nonidentical chains – α and β, connected by S-S-bonds (UniProtKB/Swiss-Prot: P00738 (HPT_HUMAN)). The three genetic types of human haptoglobins (Hp 1-1, Hp 2-1, Hp 2-2) differ in the α-chains structures (Table 1.2).

Table 1.2. Parameters of molecule and polypeptide chains of haptoglobin (по White[*] et al., 1978; Shulz, Schirmer[**], 1979)

Parameters of haptoglobin molecule and its polypeptide chains	Genetic types of haptoglobins		
	1-1	2-1	2-2
Number of α-chains MM of α- chain, kDa	2 9		
Number of α^2-цепей, MM of α^2- chain, kDa	-		2 16[*];19,8[**]
Number of β-цепей, MM of β- chain, kDa	2 42,6		
Number of S-S-связей In molecule of haptoglobin	9	10	11
Structural formula of Hp	$\alpha_2\beta_2$	$\alpha^1{}_2\beta_2\,(\alpha^2\beta)_n$	$(\alpha^2{}_2\beta_2)_n$
MM of haptoglobin, kDa	103	162, 220, 280 etc.	117, 234, 350 etc.

α^1-chain consists of 84 amino acids and exists in two modifications – α^{1F} and α^{1S}, which vary in one amino acid. This chain is a part of Hp 1-1. α^2-chain is presented in Hp 2-1 and 2-2 and has 142 amino acid residues (UniProtKB/Swiss-Prot: P00738 (HPT_HUMAN)) (Table 1.2) (Figure 1.10).

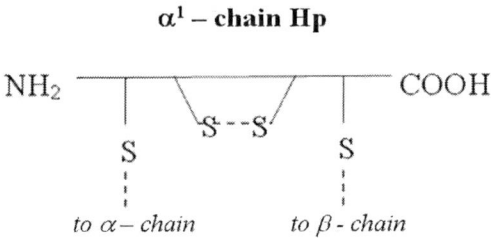

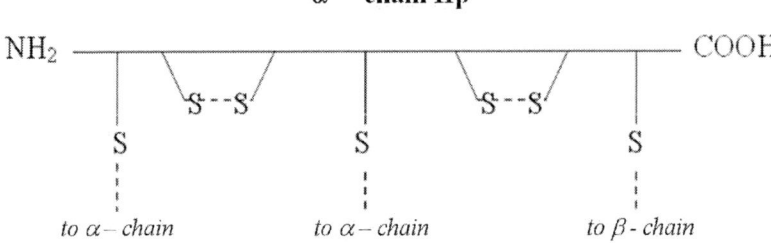

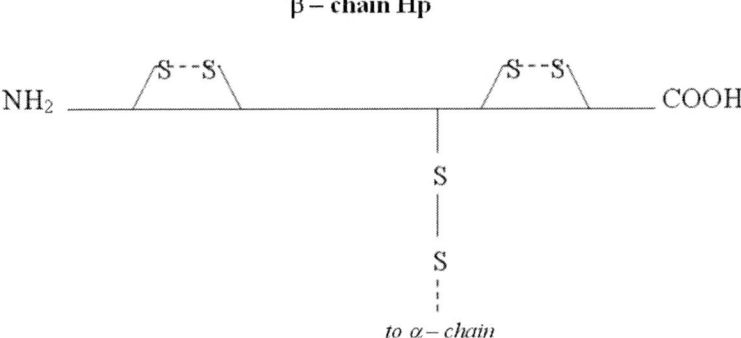

Figure 1.10. α-and β- chains of human haptoglobin.

Structural formulas of Hp 1-1, 2-1 and 2-2 are presented in the Figures 1.11, 1.12, 1.13.

Alpha and beta-chains arise from the same preproprotein, which is encoded by haptoglobin gene, located on 16th chromosome (NCBI Reference Sequence: NM_001126102.1; Goldstein, Heath, 1984). Beta-chain consists of 245 amino acids (UniProtKB/Swiss-Prot: P00738 (HPT_HUMAN)).

Thereby, haptoglobins are monomeric proteins, consisting of the polypeptide chains, bond with S-S-bonds.

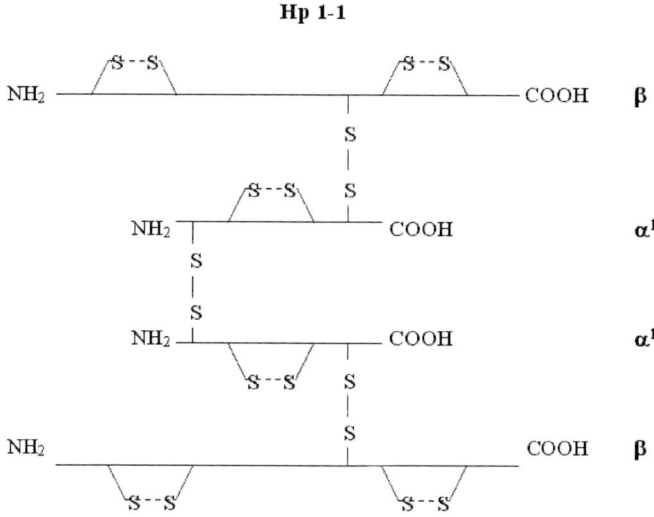

Figure 1.11. Structural formula of human haptoglobin Hp 1-1.

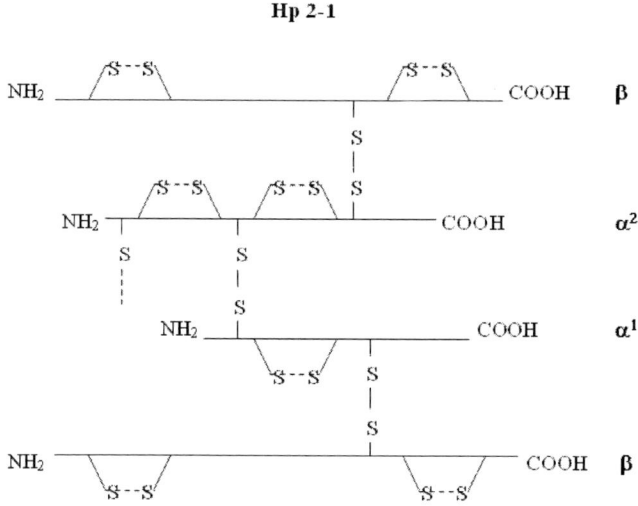

Figure 1.12. Structural formula of human haptoglobin Hp 2-1.

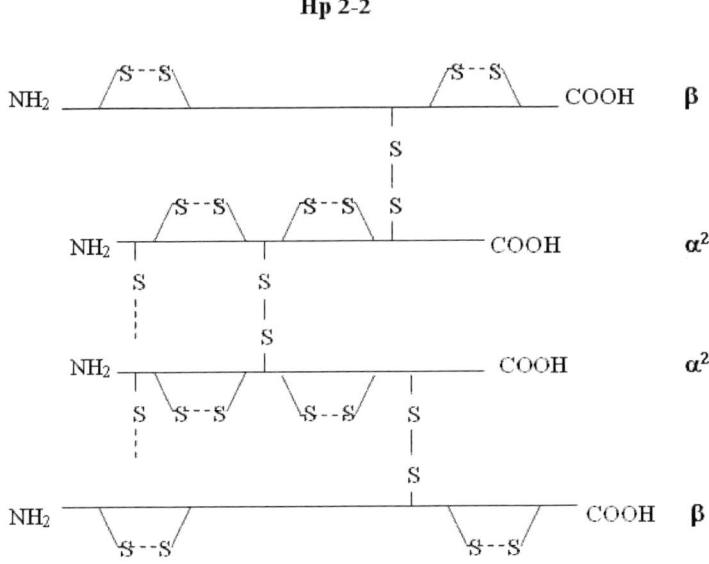

Figure 1.13. Structural formula of human haptoglobin Hp 2-2.Transferrin.

Iron concentrations in blood and other fluids of the organism are regulated by means of proteins from transferrin family, which bind and carry Fe^{3+} (family: "transferrin family" in UniProtKB; Aisen et al. 1978; Harris, Aisen 1989). The serum transferrin TFs, lactoferrin LTFs, ovotransferrin oTFs and melanotransferrin belong to this family (Woodbury et al. 1980; Wally, Buchanan, 2007). Human transferrin gene is situated on the 3rd chromosome. It is considered that transferrin gene generates as a result of ancestral gene duplication, which led to formation of homologous domains in the protein, each of them having the iron binding center (NCBI Reference Sequence: NM_001063.3).

In all mammals transferrin is monomeric protein and glycoprotein with MM about 76,5 kDa (Jeffrey et al., 1998; Wally et al., 2006; Thakurta et al., 2004; Wally, Buchanan, 2007). Polypeptide chain consists of 698 amino acids; there are two domains in its structure (Figure 1.14, 1.15).

The structure of the molecule is stabilized by 18 disulphide bonds (UniProtKB/Swiss-Prot: P02787 (TRFE_HUMAN)), the number of which varies in different proteins belonging to the family: hTF has three more of them compared to LTF (Figure 1.15) (Wally, Buchanan, 2007).

```
MRLAVGALLVCAVLGLCLAVPDKTVRWCAVSEHEATKCQSFRDHMKSVIPSDGPSVACVK
KASYLDCIRAIAANEADAVTLDAGLVYDAYLAPNNLKPVVAEFYGSKEDPQTFYYAVAVV
KKDSGFQMNQLRGKKSCHTGLGRSAGWNIPIGLLYCDLPEPRKPLEKAVANFFSGSCAPC
ADGTDFPQLCQLCPGCGCSTLNQYFGYSGAFKCLKDGAGDVAFVKHSTIFENLANKADRD
QYELLCLDNTRKPVDEYKDCHLAQVPSHTVVARSMCCKEDLIWELLNQAQEHFCKDKSKE
FQLFSSPHGKDLLFKDSAHGFLKVPPRMDAKMYLGYEYVTAIRNLREGTCPEAPTDECKP
VKWCALSHHERLKCDEWSVNSVGKIECVSAETTEDCIAKIMNGEADAMSLDGGFVYIAGK
CGLVPVLAENYNKSDNCEDTPEAGYFAIAVVKKSASDLTWDNLKGKKSCHTAVGRTAGWN
IPMGLLYNKINHCRFDEFFSEGCAPGSKKDSSLCKLCMGSGLNLCEPNNKEGYYGYTGAF
RCLVEKGDVAFVKHQTVPQNTGGKNPDPWAKNLNEKDYELLCLDGTRKPVEEYANCHLAR
APNHAVVTRKDKEACVHKILRQQQHLFGSNVTDCSGNFCLFRSETKDLLFRDDTVCLAKL
HDRNTYEKYLGEEYVKAVGNLRKCSTSSLLEACTFRRP

>sp|P02787|TRFE_HUMAN Serotransferrin OS=Homo sapiens
GN=TF PE=1 SV=3
```

Figure 1.14. Amino acid sequence of human serum transferrin (UniProtKB, BLASTP 2.2.25 [Feb-01-2011]).

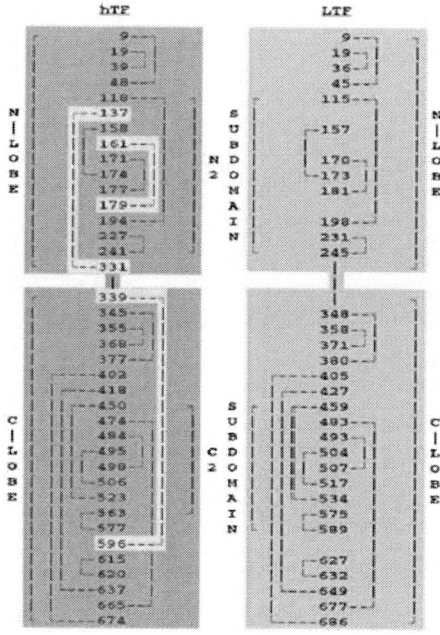

Figure 1.15. Disulfide bonds within the structure of hTF (blue) (Wally et al., 2006) and LTF (red) (Jameson et al., 1999). hTF contains three additional S-S-bonds as compared to LTF (yellow). (According to Wally, Buchanan, 2007).

Hemopexin

Another serum β-globulin hemopexin binds heme and prevents its excretion with urine, thus retaining the heme iron for further utilization. Albumin is also capable of binding heme, but its affinity with heme is lower than of hemopexin.

Human hemopexin is encoded by single gene or possibly several genes. Hemopexin gene is localized on 11th chromosome (Figure 1.16) (Source: HGNC Symbol; Acc:5171; NCBI Reference Sequence: NM_000613.2).

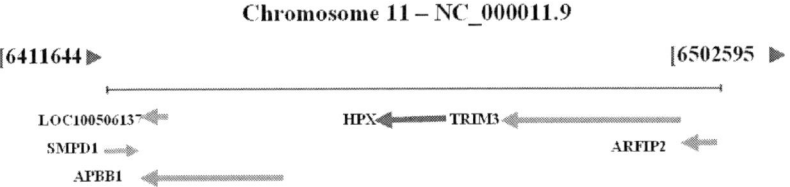

Figure 1.16. Location of human hemopexin gene on 11th chromosome.

The protein has two repeats (replicates) in its structure, which are probably the result of duplication (Altruda et al., 1985; Takashi et al., 1985). The polypeptide chain consists of 462 amino acids (UniProtKB/Swiss-Prot: P02790 (HEMO_HUMAN)). Hemopexin is a glycoprotein and monomeric protein with MM about 63 kDa, its structure is stabilized by six S-S-bonds (Takashi et al., 1985).

The similar hemopexin structure is found in rat *Rattus norvegicus* (UniProtKB/Swiss-Prot: P20059 (HEMO_RAT)), mouse *Mus musculus* (UniProtKB/Swiss-Prot: Q91X72 (HEMO_MOUSE)), bull *Bos Taurus* (UniProtKB/Swiss-Prot: Q3SZV7 (HEMO_BOVIN)), pig *Sus scrofa* (UniProtKB/Swiss-Prot: P50828 (HEMO_PIG)).

Hemoglobin

Unlike extracellular blood plasma proteins hemoglobin is the intracellular protein. Hemoglobin is the member of globin family, in which proteins with different tissue localization are included: hemoglobin, myoglobin, neuroglobin, cytoglobin and globin X, found only in fishes and amphibians (UniProtKB: family."globin family"; Weber, Vinogradov 2001; Burmester et al., 2002; Burmester, Hankeln, 2004; Freitas et al., 2004; Roesner et al., 2005).

The hemoglobin molecule consists of two alpha- and two beta-chains, which are linked one to another by noncovalent bonds to form tetramer. Therefore, the hemoglobin molecule is organized as a oligomer – tetramer. The human alpha gene cluster located on chromosome 16 spans about 30 kb and includes seven loci (Figure 1.17): $5^/$ - zeta – pseudozeta – mu – pseudoalpha-1 – alpha-2 – alpha-1 – theta – $3^/$.

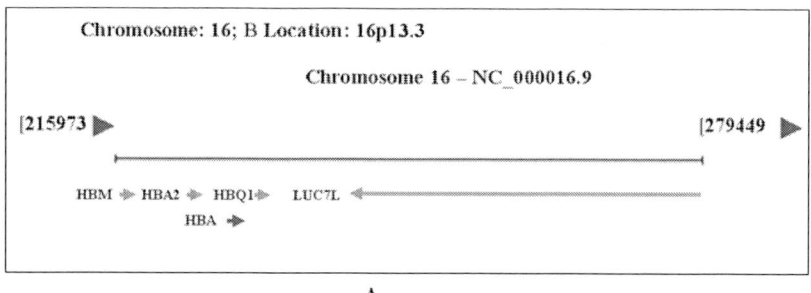

A

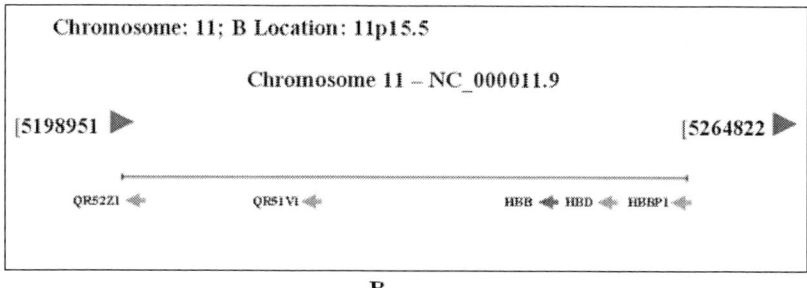

B

Figure 1.17. The human alpha (A) and beta (B) HBA and HBB genes clusters.

The alpha-2 (HBA2) and alpha-1 (HBA1) coding sequences are identical. These genes differ slightly over the $5^/$ untranslated region and the introns, but they differ significantly over the $3^/$ untranslated regions. Two alpha chains plus two beta chains constitute HbA, which in normal adult life comprises about 97% of the total hemoglobin; alpha chains combine with delta chains to constitute HbA-2, which with HbF (fetal) makes up the remaining 3% of adult hemoglobin (NCBI Reference Sequence: NM_000558.3).

Beta-chain genes are located on 11th chromosome; the order of the genes in the beta-globin cluster is 5'-epsilon -- gamma-G -- gamma-A -- delta -- beta--3'(NM_000518.4) (Figure 1.17).

1.5. THE ROLE OF OLIGOMERIC PROTEINS FORMATION AND DISSOCIATION IN OSMOREGULATION

Intracellular oligomeric proteins assemblage from monomers leads to decrease of intracellular osmotic pressure, and dissociation of oligomers leads to its increase. Since cellular membrane is impermeable even for small proteins, oligomers dissociation doesn't result in lost of subunits. The way of regulation of osmotic pressure by means of oligomer-monomer transitions is atypical for extracellular fluids, as subunits, generating in course of oligomer dissociation, filtrate through the capillary walls and excrete from the organism. For example, beta-2 microglobulin, which is a small soluble subunit of HL-A proteins, can excrete from the organism with urine (Shulz, Schirmer, 1979).

Chapter 2

THE ORGANIZATION OF BLOOD PLASMA PROTEINS IN CARTILAGINOUS FISHES *CHONDRICHTHYES*

2.1. SPECIFICITIES OF EXTRACELLULAR FLUIDS COMPOSITION IN CARTILAGINOUS FISHES ORGANISM. ARE THERE ANY ALBUMINS IN THE BLOOD OF CARTILAGINOUS FISHES?

Environment and Extracellular Fluids of the Cartilaginous Fishes

Cartilaginous fishes (*Chondrichthyes*), which diverged from primitive jaw-gilled fishes *Acanthodii* (sometimes called spiny sharks) in the beginning of Devonian, are primarily maritime dwellers.However, in spite of the fact that the majority of elasmobranch fishes are considered as the stenohaline sea forms, it is known that these fishes are successfully adapted to the fluctuations of salinity under laboratory conditions (Burger, 1965; Goldstein, Forster, 1971; Hazon et al. 1999), and such species as bull shark *Carcharhinus leucas* and Atlantic Stingray *Dasyatis sabina* populate both the sea and fresh waters (Thorson et al, 1973; Piermarini, Evans, 1998).It is show that elasmobranch fishespossess (in a different degree) a physiological ability to survive with the changing salinity of water (Anderson et al, 2002).

Their internal fluid environmentis characterized by high concentrations of urea (0.19-0.6M) and trimethylaminoxid TMAO (to 0.07-0.1M and above). Urea penetrates into organism cells by means of passive diffusion with no active mechanisms involved (Walsh et al., 1994). Such composition of organism fluids of these fishes allowed them to raise osmotic pressure of extracellular fluids up to slightly hypertonic level compared to see water, in spite of low salt concentrations (1.42-1.77%) (Anderson et al., 2002, 2007; Pillans, Franklin, 2004; Speers-Roesch et al., 2006; Villalobos, Renfro, 2007).

About the Effect of Urea and TMAO on the Blood Plasma Protein Repertoire in Cartilaginous Fishes

High urea concentrations in elasmobranches blood were advanced as an argument in support of the suggestion about the absence of albumins in it: as osmotically active non-protein compounds probably would make the presence of low-molecular osmotically active proteins in blood unnecessary (Fellows, Hird, 1981; Peters, Davidson, 1991). There is also an extra argument in support of this suggestion – the discovery of damaging effect of high urea concentrations on proteins (Lee et al., 1991).

TMAO "Protects" the Blood Proteins from the Damaging Effect of Urea

Despite the damaging effect of high urea concentrations on proteins, some organisms accumulate it even when proteins function is endangered (Wang et al., 1999). It appeared that there are such compounds in fish blood which protect the proteins from damage – they are methylamines, including TMAO, and the urea "harm" is compensated not only by its "useful" osmotic properties, but also by its cardio protective effect (Cordier et al., 1957). Thus, urea and TMAO do not interfere with albumins in cartilaginous fish blood, but mutually complement each other – urea is a great osmotic regulator and TMAO protects the proteins from its damaging action.

In which Fishes the Urea Content in Extracellular Fluids of the Organism Is as High as in Sharks? Are There any Albumins in These Fishes' Blood?

Just like elasmobranches, cyclostomes, fringe-finned (or lobe-finned) fishes (*Crossopterygii*) and their specialized branch – dipnoans (*Dipnoi*) - also have high concentrations of urea in blood and tissue fluid (Lutz, Robertson, 1971, Hyodo et al., 2007). In case of coelacanths it is determined that urea in their organisms is used the same way as in sharks and skates, for the osmoregulation (Brown, Brown, 1967). It is believed that there are no albumins in these fishes blood. However, later not only albumin-like proteins were found in cyclostomes, in fringe-finned fishes and in dipnoans, but the proteins were found, which are very similar to mammalian albumins (Drilhon, Fine, 1959; Rall et al., 1961; Griffith et al., 1974; Sulya et al., 1961; Manwell, 1963; Masseyeff et al., 1963; Metcalf et al., 2007). And in coelacanths the blood albumin content appeared to be low (Metcalf et al., 2003). Australian lungfish *Neoceratodus forsteri*albumin proved to be similar to albumin of the mammals (Metcalf et al., 2007). Its NH_2-terminal fragment of 101 amino acids had high identity level with the same fragment of mammalian albumin. Similar to human albumin, the albumin from *Neoceratodus forsteri* bound [^{14}C]-palmitic acid and didn't contain carbohydrate in its molecule structure (Metcalf et al., 2007).

2.2. SERUM GLOBULINS FROM SHARKS AND RAYS

Immunoglobulins

Cartilaginous fishes are the most primitive species in which the ability to synthesize antibodies was found (Mestel, 1996). Their immunoglobulins belong to three types: M, W and NAR; besides the immunoglobulin-like protein is found, which is capable of binding L-chains in particular way (Ota et al., 2003; Hsu et al., 2006; Ohta, Flajnik, 2006). Unlike mammalian, the heterogeneity of cartilaginous fishes serum antibodies is expressed poorly, and their repertoire doesn't change after second antigen introduction (Flajnik, 1996).

Immunoglobulin genes in cartilaginous fishes are organized in clusters. Each cluster has on the one V, J and C segment, and there can be several tens or hundreds of such clusters in a locus (Litman et al., 1984). The diversity of antibodies in cartilaginous fishes is formed not by combinatorial combination of different segments, but by expression of large amount of clusters. There are three types of L-chain genes organization, which differ in the way of joining of segments in clusters: in the 1st and 3rd type loci - V, J and C segments are separated, and in the 2nd type loci – V and J segments are already joined in germinal genome (Figure 2.1) (Rast et al., 1994).

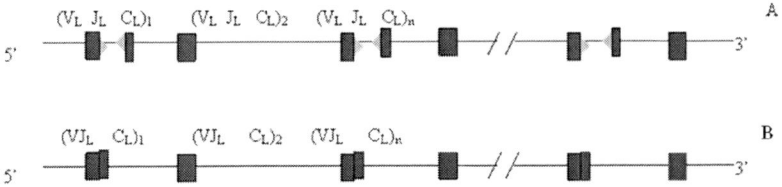

Figure 2.1.Scheme of the structure of the loci of IGL of the shark(according toRast et al., 1994). A and B – different types of organization IGL (explanation in text).

The genes of all three types have low degree of homology among them. The 1st genes type includes the genes of horn shark *Heterodontus francisci* (Shamblott, Litman, 1989), little skate *Leucoraja erinacea* (Raja erinacea) (Anderson et al., 1995); the 2nd type includes the genes of the spotted ratfish *Hydrolagus colliei* (Maisey, 1984), the sandbar shark *Carcharhinus plumbeus* (Hohman et al., 1992; Hohman et al., 1993); and the 3rd type includes the genes of nurse shark *Gynglimostoma cirratum* (Greenberg et al., 1993) and horn shark (Rast et al., 1994).

The analysis of organization of immunoglobulin H-chain genic segments was carried out for the horn shark *Heterodontus francisci*(UniProtKB/Swiss-Prot: P03983-PO3988, P23084-P23088, P83907, P83742, P83743). Amino acidsequences are determined for heavy chain V region of immunoglobulin of horn shark *Heterodontus francisci* (Figure 2.2, 2.3), IGH of spiny dogfish *Squalus acanthias* (Bartl, Weissman, 1994), smooth dogfishs *Mustelus canis* (Marchalonis, Edelman, 1966), sting ray (Marchalonis, Schonfeld, 1970), nurse shark *Gynglimostoma cirratum* (UniProtKB/Swiss-Prot: P83977, P83984, P83985, Q90523), shovelnose guitarfish *Rhinobatos productus* (Rumfelt et al., 2004). Amino acid sequencesof other proteinsof Igsuperfamily was made in sandbar shark *Carcharhinus plumbeus* (*Squalus plumbeus*) (UniProtKB/Swiss-Prot: D0EP38-D0EP40), lemon shark *Negaprion brevirostris* (*Hypoprion brevirostris*) (UniProtKB/Swiss-Prot: Q56II0), spiny dogfish *Squalus acanthias* (UniProtKB/Swiss-Prot: D5FGJ5-D5FGJ57), spotted wobbegong *Orectolobus maculates* (UniProtKB/Swiss-Prot: A9CBG4, A9CBG5), clearnose skate *Raja eglanteria* (UniProtKB/Swiss-Prot: Q8AXA0), pacific electric ray *Torpedo californica* (UniProtKB/Swiss-Prot: Q07153), little skate *Leucoraja erinacea* (*Raja erinacea*) (UniProtKB/Swiss-Prot: D5FGF4-D5FGF6 and D5FGF8), cownose ray *Rhinoptera bonasus* (UniProtKB/Swiss-Prot: Q56II2), horn shark *Heterodontus francisci* (UniProtKB/Swiss-Prot: 20938).

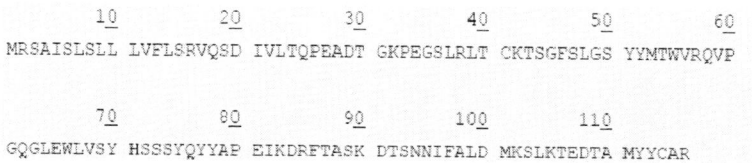

```
        10         20         30         40         50         60
MRSAISLSLL LVFLSRVQSD IVLTQPEADT GKPEGSLRLT CKTSGFSLGS YYMTWVRQVP

        70         80         90        100        110
GQGLEWLVSY HSSSYQYYAP EIKDRFTASK DTSNNIFALD MKSLKTEDTA MYYCAR
```

Figure 2.2. Amino acid sequence (complete) of immunoglobulin heavy chain V region of Horn shark *Heterodontus francisci*. Sequence length 116 AA. UniProtKB/Swiss-Prot: P03983.

```
        10         20         30         40         50         60
ATPSPAILYG LCSCEQTNTD GSLAYGCLAM DYSPEITSIT WKKDKEPITT GLKIYPSVLN

        70         80         90        100        110        120
KKGTYTRSSQ LTITESEVGS SKIYCEVRRG ESLWIKEILD CKGDIVFPTV ILTQSSSEEI

       130        140        150        160        170        180
TSRRFATVLC SIIDFHPESI TVSWLKDGQP MDSGFVTSPT CEVNGNFSAT SRLTVPAGEW

       190        200        210        220        230        240
FSNTVYTCQV AHQEVTQSRN ITGSQVPCSI GDPVIKLLPP SIEQVLLEAT VTLTCVVSNA

       250        260        270        280        290        300
PYGVNVSWTQ EKKPLKSEIA VQPGEDSDSV ISTVNISTQA WLSGAEFYCV VSHQDLPTPL

       310        320        330        340        350        360
RASIHKEEVK DLREPFVSVL LFPAEDVSAQ RFLSLTCLVR GFSPREIFIK WTVNDKSVNP

       370        380        390        400        410        420
GNYKNTEVMA ENDNRSFFIY SLLSIAAESW ASGASYSCVV GHEAIPLKII NRTVNKSSGK

       430
PSFVNISLAL LDTVNSCQ
```

Figure 2.3. Amino acid sequence (fragment, clone 12022) of Ig heavy chain C region of Horn shark *Heterodontus francisci*. Sequence length 438 AA. UniProtKB/Swiss-Prot: P23085.

All proteinsofIgsuperfamily have Y-shaped structure; they are glykoproteins and are organized as monomeric proteins. MM values for immunoglobulins from the smooth dogfishs *Mustelus canis* amounted to 198 and 982 kDa (Marchalonis, Edelman, 1966). μ-like heavy chain with MM about 63 kDa is identified in the IgM (Ota et al., 2003). In the Y-shaped serum proteins of buckler skate (thornback ray) *Raja clavata* and blue stingray (common stingray) *Dasyatis pastinaca* the chain diversity was discovered with MM from 24 to 250 kDa (Figure 2.4).

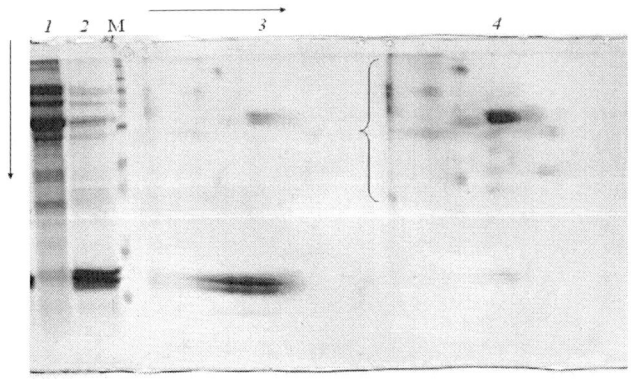

Figure 2.4. 2D-SDS-electrophoresis of serum proteins of common stingray *Dasyatis pastinaca* L. Vertical pointer show the direction of SDS-electrophoresis, horizontal pointer – of disk-electrophoresis; curly bracket show L and H Ig chains. *1, 4* – serum, *2, 3* – peritoneal fluid.

Hemopexin, Transferrin and Haptoglobin

In the beta-globulin fraction of the blood serum from spiny dogfish *Squalus acanthias* the protein, which binds hemin, was found (Figure 2.5) (Andreeva, 1997). With help of the Muller reagent (Palmour, Sutton, 1971) the protein with MM about 80 kDa (concentration PAGE gradient, non-denaturing conditions), capable of binding Fe^{3+}, was revealed in the same fraction. Electrophoretic mobility R_f of this protein in disk electrophoresis is 0,33 (Figure 2.5). The proteins were identified as hemopexin and transferrin respectively. The comparison of MM values of these proteins under the concentrations PAGE gradient, in PAGE with urea and SDS-PAGE (reducing conditions) showed that they were monomers (Andreeva, 2008).

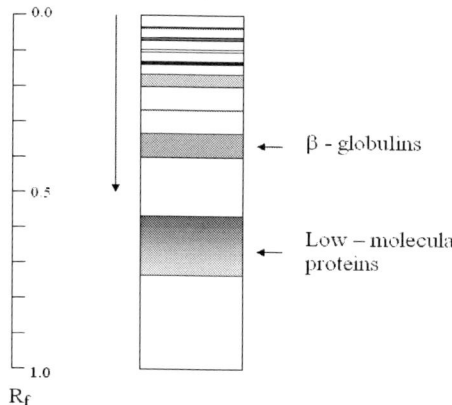

Figure 2.5. Scheme of disk-electrophoresis of serum proteins of spiny dogfish *Squalus acanthias*. Vertical pointer show the direction of electrophoresis, horizontal pointers indicate to b-globulins and low-molecular proteins. R_f - electrophoretic mobility. (Andreeva, 2001a).

```
         10         20         30         40         50         60
    MKFLLGGLCL FWAVALSFCF PWLKHQPNIT EDELEQHHSH PRHEGFPDRC DGLGFDAVTL

         70         80         90        100        110        120
    PEQSVTYFFR DEFLWRGERF AAEFINKTWP GLPDHIDAAP PIHHKNSPEQ HDPMFFFKGN

        130        140        150        160        170        180
    QVWQYYGTKL EQQFLIQDKF HOIPDNLGAA VECPEGECQH DSVLFFKGAI TYVFDLSTMT

        190        200        210        220        230        240
    VKQRPWTGVH ICTAAMRWID RYYCFQGSNT TRFYPHTGLV TENYPKDAFN YTMRCZGRGH

        250        260        270        280        290        300
    GNKTVDPSIH NRCSNRSFDE FNQDEFGRVY AFRAGWYFRL DSKRDGWHAM PIHSTWPSLH

        310        320        330        340        350        360
    GKIDGVFSWN KKTYFIQGSQ IYIYKAEAHY TLIENYPKSV TESDSIHSTD VDATFICPGT

        370        380        390        400        410        420
    SILHVISGNQ VQSINLEQTP RNLVDGLRIG HSHVDGAMCN SRGIFIFVGT DYYKYPSTTE

        430        440
    LAGWTKIPEP HSIRADFMSC VQ
```

Figure 2.6. Amino acid sequence (complete) of hemopexin of Nurse shark *Ginglymostoma cirratum*. Sequence length 442 AA. UniProtKB/TrEMBL: E7CQA0 (E7CQA0_GINCI).

The hemopexin polypeptide chain from nurse shark *Ginglymostoma cirratum* consists of 442 amino acid residues(Figure 2.6) (UniProtKB/TrEMBL: E7CQA0), from little skate *Leucoraja erinacea* (*Raja erinacea*) – of 437 (Figure 2.7) (UniProtKB/TrEMBL: E7CQ99), from cloudy catshark *Scyliorhinus torazame* – of 613 amino acid residues (UniProtKB/TrEMBL: Q9I9J7).

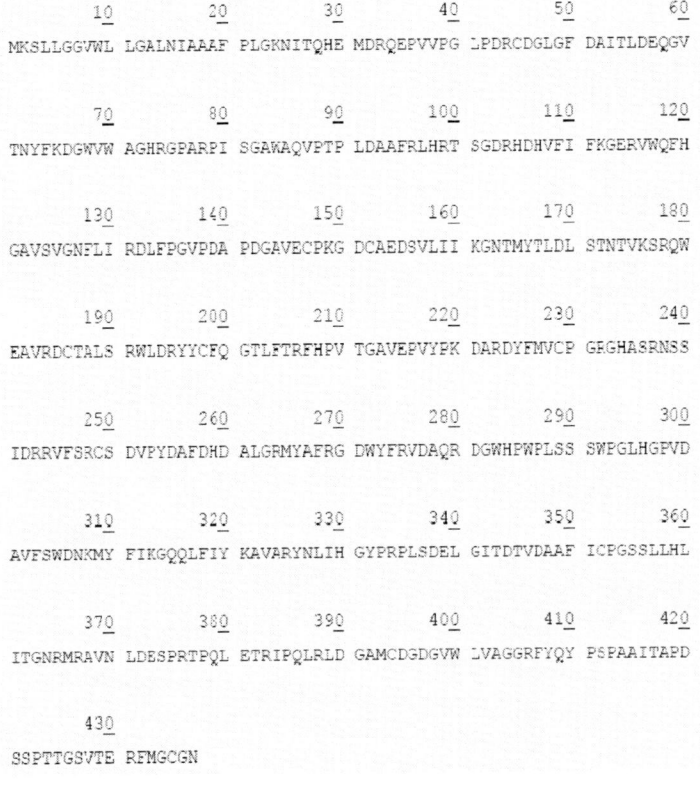

Figure 2.7. Amino acid sequence (complete) of hemopexin from little skate *Leucoraja erinacea* (*Raja erinacea*). Sequence length 437 AA.UniProtKB/TrEMBL: E7CQ99 (E7CQ99_LEUER).

There is no evidence of haptoglobin presence in cartilaginous fishes blood; in the liver of little skate the protein resembling haptoglobin with polypeptide chain, consisting of 432 amino acids, was found (Figure 2.8) (UniProtKB/TrEMBL: Q98983).

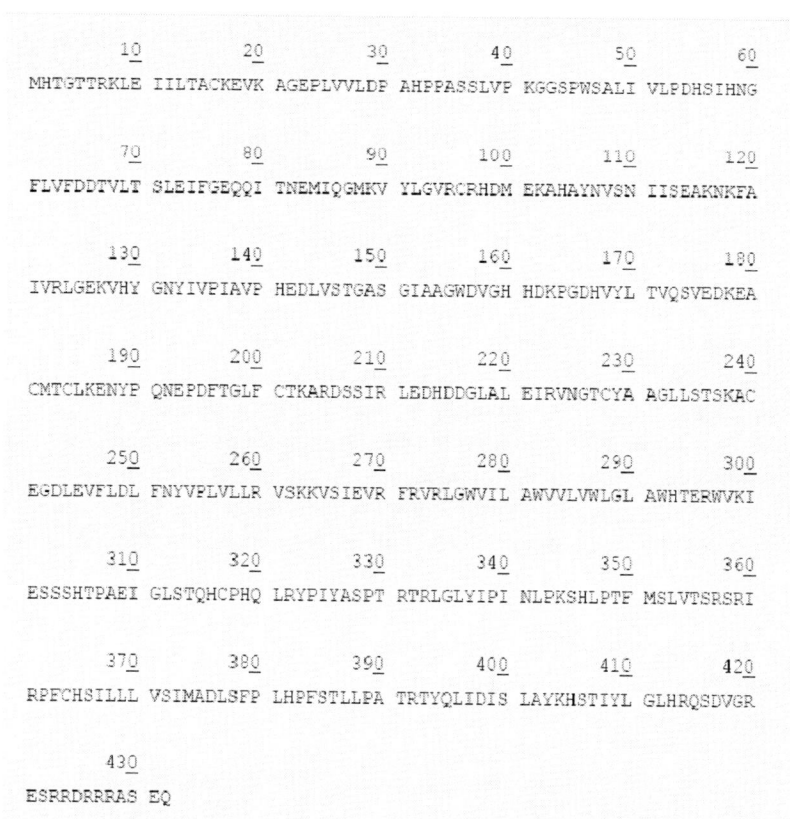

Figure 2.8. Amino acid sequence (complete) of haptoglobin-like protein from little skate *Leucoraja erinacea* (*Raja erinacea*). Sequence length 432 AA. UniProtKB/TrEMBL: Q98983 (Q98983_LEUER).

2.3. LOW-MOLECULAR BLOOD PLASMA PROTEINS IN SHARKS AND SKATES

The Content of Low-Molecular Proteins in Blood and Their Structural Features

As far back as half a century ago Cordier (Cordier et al., 1957) revealed an albumin-like protein in the blood serum from small-spotted catshark *Scyliorhinus canicula* in concentration 0.4% and 37% from the total protein, which was 1.07%. Albumin was found in shark and scate fry, inhabiting river mouths. It wasn't found in mature fish; among immature fish, caught in the sea, the specimens with intermediate proteinogram type were found (Saito, 1957). Albumin was also found in the plasma from bonnethead shark or shovelhead *Sphyrna tiburo* in concentrations from 0.3 to 0.5%, or about 13.5% from the total plasma protein, which was from 2.2 to 4.3% (Harms et al., 2002). Two low-molecular proteins from the blood serum of the smooth dogfishs *Mustelus canis* were described – C-reactive protein and P-component, having homologous NH_2-fragments of 20-amino acids each, however there is no information whether these proteins belong to albumins (Robey et al., 1983). C-component appeared to be a simple protein with MM about 50 kDa, capable of binding Ca^{2+}; and P-component appeared to be a glycoprotein with MM about 25 kDa. In the blood from the sandbar shark *Carcharhinus plumbeus* a low-molecular protein was revealed, consisting of two subunits (Vazquez-Moreno et al., 1992), which had similar NH_2-terminal fragments; there is no data about the resemblance of this protein to albumins from higher vertebrates.

The Organization of Low-Molecular Blood Plasma Proteins in Shark Spiny Dogfish *Squalus Acanthias* L

Multiple low-molecular proteins were found in blood serum of the shark *Squalus acanthias* L., which was caught in the Black Sea (Figure 2.4) (Andreeva, 1986b). Most mobile fraction of the serum, which made up to 50% of the total protein (Figure 2.5), was divided into 11-12 components in the concentration PAGE gradient, three of them had MM about 58-60 kDa, six – 64-70 kDa, and three – 130-134 kDa (Figure 2.9). Only one component withMM about 45 kDafrom most mobile fraction was revealed in PAGE with 8M urea and in SDS-PAGE(Figure 2.9).

The observed variety can be due to addition of different ligands to the protein molecule with MM about 45 kDa by means of non-covalent bonds, that modulates its charge and MM value. Among those ligands are lipids and carbohydrates, including those carbohydrates, covalently bonded with the protein. Low-molecular proteins from spiny dogfishbonded hemin and also, like mammalian albumin, albumin-specific dye Evans blue.

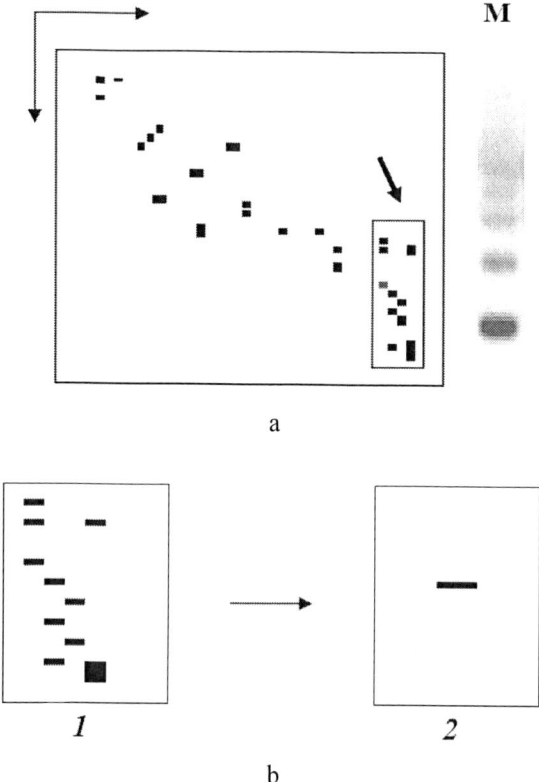

Figure 2.9. Scheme of electrophoresis of serum proteins of spiny dogfish *Squalus acanthias* in concentration PAGE gradient (3-20%) (A, B1) and in SDS-PAGE (B2). M – BSA. Vertical pointer show the direction of concentration PAGE gradient electrophoresis, horizontal pointer – of disk-electrophoresis. Small pointer indicates to low-molecular proteins (A, B1), which are presented by one band in the SDS-electrophoresis (B2). (Andreeva, 2008).

The Organization of Low-Molecular Blood Plasma Proteins from Buckler Skate *Raja ClavataL.* and from Common Stingray *Dasyatis Pastinaca* L

The heterogenic low-molecular fraction (LMF) was revealed in the blood plasma and interstitial fluid of buckler skate; it consisted of 4-5 proteins, which had the highest electrophoretic mobility values in the disk electrophoresis. Two proteins from this fraction (MM ranging from 62 to 80 kDa) made about 56% of total blood serum protein, and other proteins from this fraction (MM ranging from 20 to 40 kDa) made about 4,3%.

The heterogenic LMF of the stingray blood serum consisted of 8 proteins. Two macrocomponents with MM about 67 and 47 kDa and the subfraction of six proteins with MM about from 13 to 30 kDa (marked as *fast*-fraction) were included into this fraction (Figure 2.10). Relative content of low-molecular proteins in the serum did not exceed 2.9% of the total protein; meanwhile, it reached 28.4% in interstitial fluid of the muscles, mainly by the component with MM about 47 kDa (it made up to 20.5% of all muscle fluid proteins). Addition of 8M urea to the serum led, firstly, to dissociation of proteins with MM 120 and 200 kDa to the proteins with MM about 34 kDa and to the replenishment of LMF with them; and secondly, to reduction of *fast*-fraction heterogeneity from six components to one with MM about 13 kDa (Figure 2.10) (Andreeva, Fedorov, 2010).

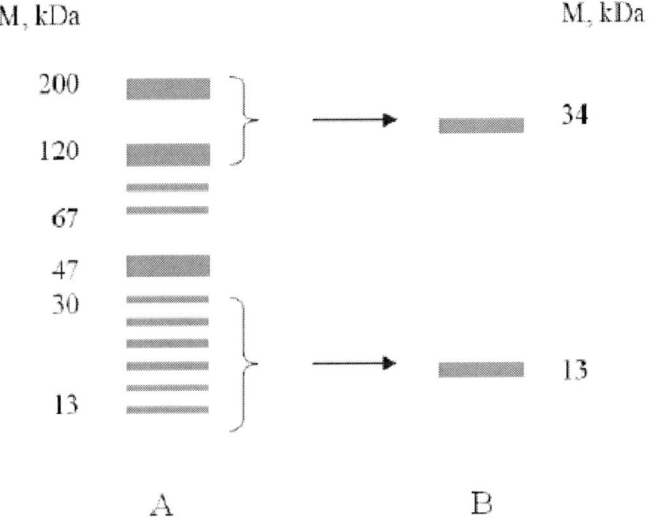

Figure 2.10. The change of heterogeneity and qualitative composition of serum proteins with MM in range 13 - 200 kDa in electrophoresis: in concentration PAGE gradient (5-40%) (A) and in PAGE with urea (B), scheme.

2.4. DIFFERENT BEHAVIOR OF THE LOW-MOLECULAR BLOOD PLASMA PROTEINS OF ELASMOBRANCHES IN THE UREA-CONTAINING AND UREA-FREE ENVIRONMENT

Different blood plasma proteins of cartilaginous fishes behave differently in urea-containing solutions: in the leafscale gulper shark *Centrophorus squamosus* (Bonnaterre) the differentiation of lipoproteins of low- and high-density by solubility in urea solution *in vitro* is discovered (Mills et al., 1977).

For other fish species the effect of urea on the proteins aggregation degree is described (Andreeva, Fedorov, 2010).For example, serum gamma-2-globulins (including immunoglobulins) from the spiny dogfish and skates *Raja clavata* L. and *Dasyatis pastinaca* L. remained at the starting zone in 2D-electrophoresis with urea and without it. In the same way, transferrins had the same MM values – about 80 kDa – in 2D-PAGE-electrophoresis with urea and without it. It follows from this that aggregation into protein complexes without urea and dissociation of such complexes under action of urea wasn't observed *in vitro* for these groups of proteins.Low-molecular fractions behaved quite differently. For the common stingray, the transformations of low-molecular proteins, described in the previous part, can be explained by their aggregation into the complexes with MM 120 and 200 kDa without urea and dissociation of these complexes into small proteins by hydrogen bonds breakage under the urea action. Heterogeneity reduction in *fast*-fraction under the presence of urea can be explained by the fact that all the components of this fraction are formed by oligomerization or covalent modification of the single protein (Andreeva, Fedorov, 2010).

The detection of the proteins, capable to aggregation and dissociation depending of the presence of urea in medium, in blood and interstitial fluid of common stingray and spiny dogfish, made it possible to extend the conceptions about low-molecular proteins organizations in this group of fishes. The comparison of these proteins with low-molecular proteins of the fishes from other groups and mammals has revealed. The following differences in the way of organization, MM values and surface structure of proteins were revealed:

1) in the presence of urea low-molecular proteins of common stingray are presented by several monomeric proteins, which aggregate into complexes without urea. Similar proteins were also found in the blood of fresh-water bony fishes: in the presence of urea they dissociated into 10-13 low-molecular subunits (Andreeva, 1999). Unlike them, mammalian albumins are monomers,

and the presence of urea in reaction medium does not change their organization.

2) LMF of the common stingray serum was presented by 8 proteins with MM from 13 to 67 kDa without urea, and only by two types of molecules with MM 34 and 13 kDa in the presence of urea. These protein MM values did not coincide with values of the low-molecular proteins from other cartilaginous and teleost species: thus, for the smooth dogfishs low-molecular proteins with MM 50 and 25 kDa are described (Robey et al.,1983); for the spiny dogfish– 12 proteins with MM from 58 to 70 kDa, which were presented as one protein with MM about 45 kDa in SDS-PAGE (Andreeva, 1999). Oligomeric proteins of the bream *Abramis brama* L. and the roach *Rutilus rutilus* L. dissociated into 10-13 proteins with MM from 18.5 to 73 kDa in the SDS-PAGE (Andreeva, 2010a, b). For the sea teleost fishes the LMF are described, consisting of 10 proteins for arctic flounder *Liopsetta glacialis* P. and of 7 proteins for shorthorn sculpin *Myoxocephalus scorpius* L. with MM from 30 to 90 kDa, of 4-5 proteins for atlantic cod *Gadus morhua* L. with MM from 45 to 80 kDa (Andreeva, 2008). Meanwhile, denatured albumin molecules from sturgeons and from mammals had similar MM values – about 67 kDa (Andreeva, 1999).The presented data show a high level of MM variability of the low-molecular proteins in fish.

3) distinctive characteristics of the surface structure of bloodproteins from skate, spiny dogfish and other fishes and higher vertebrates are manifested in different behavior of the proteins with and without urea, which can promote both dissociation and aggregation of the proteins (Alexandrov, 1985). Low-molecular proteins from the common stingray form aggregates without urea by means of hydrogen bonds, that distinguishes them from human and bovine albumins, which generate polymeric forms by means of covalent S-S-bonds (Andreeva, 2008). Therefore, the presence of urea in the reaction medium is enough to break the protein aggregates in skate, as for dissociation of polymeric forms of HSA and BSA the reducing conditions are needed too. Some sharks have proteins, stabilized by S-S-bonds also: the protein with MM 70 kDa in *Carcharhinus plumbeus* consists of two chains with MM 36 and 24 kDa, connected by S-S-bonds (Vazquez-Moreno et al., 1992); C-reactive protein with MM 250 kDa from *Mustelus canis* is presented in dimers (MM about 50 kDa) without urea, monomers of which (MM about 25 kDa) being connected by S-S-bonds, and P-component has MM 250 kDa and consists of monomers with MM 25 kDa (Robey et al., 1983). Whereas, low-molecular proteins from spiny dogfish blood, like mammalian albumins, bind albumin-specific dye Evans blue.

Based on the examples it was sugested that there is a certain similarity in the low-molecular proteins' organization in cartilaginous fishes, determined by the presence of high urea concentrations in the internal fluids of their organisms: in the urea-free medium their low-molecular proteins aggregate into high-molecular complexes, which dissociate into low-molecular proteins after addition of 8M urea. This ability of proteins to transform their structure was probably the reason, due to which low-molecular proteins could not be revealed in the blood of sharks: instead of them, the high-molecular complexes were found at the electrophoregram. It is unclear whether the reversible monomer-oligomer structural transformations of proteins are of any physiological importance, for example, in regulation of colloid-osmotic pressure of shark blood. However, it is established that the fluctuations of the water salinity cause the fluctuations of urea concentration in the extracellular fluids of the cartilaginous fishes (Anderson et al., 2007). Thus, in small-spotted catshark or lesser spotted dogfish *Scyliorhinus canicula* and banded houndshark *Triakis scyllia* the increase of water salinity leads first to an increase of Na^+ and Cl^- concentrations in the plasma and then of ureas by decreasing of its excretion and/or by increase of its production by the liver (Goldstein, Forster, 1971; Armour et al., 1993). But the fluctuations of urea concentration theoretically can provoke the reversible monomer- oligomer structural transformationsof serum proteins (Andreeva, Fedorov, 2010).

Chapter 3

CHONDROSTEI BLOOD PLASMA PROTEIN ORGANIZATION

3.1. HABITAT AND THE COMPOSITION OF ORGANISM EXTRACELLULAR FLUIDS FROM *ACIPENSERIFORMES*

The superorder *Chondrostei* consists of one order *Acipenseriformes* only, that originates from the ancient representatives of *Paleonisci*, which are related to actinopterygian fishes *Actinopterygii*. The contemporary *Acipenseriformes* are represented by the families *Polyodontidae* and *Acipenseridae*. They are morphologically strongly isolated from the others now living *Actinopterygii* (Yakovlev, 1977). Blood proteins of *Acipenseriformes* are organized in a particular manner also, they are very similar to mammals plasma proteins in their properties.

Such species as sterlet *Acipenser ruthenus*, beluga *Huso huso*, kaluga *Huso dauricus*, starred sturgeon *Acipenser stellatus*, sturgeon *Acipenser guldenstadti*, paddlefish *Polyodon spathula*, switchtail *Scaphirhynchus platorhynchus* and ship sturgeon *Acipenser nudiventris* belong to *Acipenseriformes*. They include freshwater, brackish water and migratory anadromous fishes, which inhabit fresh waters and seas with the salinity approximately 12-14‰ (the Black Sea, the Sea of Azov, the Caspian Sea) in the Russia territory.

The organism internal environment of *Acipenseriformes* fishes contains TMAO and urea (3 mM) less, and salts more (more than 1.7%) than that in cartilaginous fishes. At this the urea and salts concentrations vary significantly (Stroganov, 1962; Shilov, 1985).

3.2. SERUM GLOBULINS OF *ACIPENSERIFORMES*

Albumins andsome globulins – transferrins and immunoglobulins -are found in the sturgeon blood plasma; such globulinsas haptoglobin and hemopexin are not found in the blood plasma (Figure 3.1) (Andreeva, 1999, 2001a, 2001b).

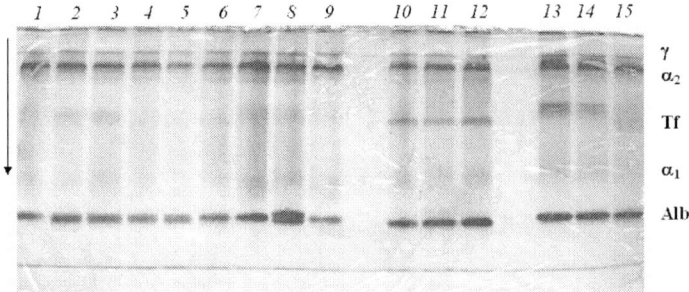

Figure 3.1. Disk-electrophoresis of serum proteins of sterlet *Acipenser ruthenus* (*1-9*), stellate sturgeon *A.stellatus* (*10-12*) and beluga *Huso huso* (*13-15*). Vertical arrow shows the direction of electrophoresis. γ -γ–globulins,α -α-globulins, Tf – transferrin, Alb – albumin.

Transferrins

In fishes and other vertebrates transferrin is the monomeric protein with MM approximately 70 kDa (Valenta et al., 1976). The majority of the fish species has one transferrin locus. The number of transferrin alleles reaches 6 for sturgeon (Balakhnin et al., 1972; Chikhachev, 1983), 3 for starred sturgeon (Chikhachev, Tsvetnenko, 1979a, b), 2 for sterlet and 3 for beluga (Chikhachev, 1983). The proteins heterogeneity in electrophoresis of more than three components is determined by sialic acids (Kirpichnikov, 1987).

In Sturgeon, Russian sturgeon, white sturgeon and sterlet the transferrins were revealed as 1-4 bands in the disk- electrophoresisby using the reagent containing nitroso-R-salt (Andreeva, 1987a, b, 1997). The formation of complete transferrin electrophoretic spectra of *Acipenseriformes* ended in the postembryonal period (Subbotkin, Subbotkina, 2003, 2004, 2011).

The purified transferrin was obtained from sterlet blood serum by the rivanol method (Palmour, Sutton, 1971) and column chromatography

(Andreeva, 1987a, b). The purified transferrin was concentrated and stabilized by salting-out in ammonium sulfate solution (Andreeva, 1997). Its solution had a specific pink color and birefringence when transferrin suspension was shaked. The birefringence indicates to the formation of transferrin microcrystals. The purified transferrin of sterlet was represented in SDS-PAGE as one band, its molecular weight was approximately 76 kDa and λ_{max}=408,3 nm. The immunological control confirmed the presence of purified transferrin in the probe (Subbotkin, Subbotkina, 2003).

The sterlet purified transferrin was used for the determination of sialic acids in its composition. The treatment of transferrin by the neuraminidase from comma bacillus (*Vibrio cholerase*) makes it possible to remove the transferrin heterogeneity. This enzyme removes sialic acids from glycoprotein. Theactivity of neuraminidase was 5 units/ 1 mg. Only one electrophoretic component remains in PAGE after treatment of purified transferrin by neuraminidase. Meanwhile mammalian transferrins are workable by more active neuraminidase - about 188 units/ 1 mg (Kirpichnikov, 1987). The calculated content of sialic acids in sterlet transferrin did not exceed 6,73M per 1M of protein (Andreeva, 1987a, b).

α-2-Neuroaminoglycoprotein

Glycoprotein with MM approximately 200 kDa was discovered in the sterlet blood serum. The relative content of sialic acids in this protein was considerably higher than in transferrin. After treatment of sterlet serum by comma bacillus neuraminidase (activity 5 units/ 1 mg) this glycoprotein disappeared from the electrophoregram because of considerable reduction in MM value after the splitting of sialic acids from it (Andreeva, 1997).

Immunoglobulins

The immunoglobulins of cartilaginous ganoids, as of the majority of bony fishes, are differentiated into the classes M and G (Marchalonis, Schonfeld, 1970; Kobayashi et al, 1982; Wang, Liu, 2007). H-chains of *Chondrostei*immunoglobulins are representedby µ-similar subunits with MM approximately 70 kDa, the information about differentiation of H-chains is not numerous(Litman et al, 1971).

It is found that sterlet L-chains are of kappa-type and IGL genes organization is segmentary, similar to the organization of mammalian kappa locus. It is proved by quantitative superiority of VL genes over CL genes in the IGL locus. So, the transition from the cluster organization of IG genes to the segmental one, typical for all mammals, occurred during the evolution of *Pisces* in *Chondrostei* fishes (Denisov, 1997; NCBI, Protein, GenBank: CAB37327-CAB37329, CAB37927, AAD22097, AAD22198, AAD22263, AAD22489; NCBI, Nucleotide, GenBank: AF129436, AF129437, AF131056). The similarity of the VL genes of *Acipenseriformes* (Siberian sturgeon) to VL ones of bony fishes, and of the CL genes of *Acipenseriformes* to CL genes of sharks was marked by a number of signs (Lundqvist et al, 1996).

Immunoglobulins and other proteins of Ig superfamily, for example, beta-2-mikroglobulin from Siberian sturgeon composed of 122 amino-acid residues (Wuertz et al, 2007; UniProtKB/Swiss-Prot: Q9PRF8) and from sterlet (fragment of 102 amino-acid residues) (UniProtKB/Swiss-Prot: Q27YD9) - are glycoproteins and are organized according to the type of monomer Y-junction molecules.

The subunits with MM approximately 23, 56, 60, 65 and 70 kDa are found in sterlet serum 2D-SDS-electrophoresis on the gamma-2-globulins track. These subunits enter to the composition of proteins with the Y-junction structure, whose MM under the nondenaturing conditions was about 890, 950 kDa and higher (high-molecular starting protein) (Figure 3.2, 3.3) (Andreeva, 2001b).

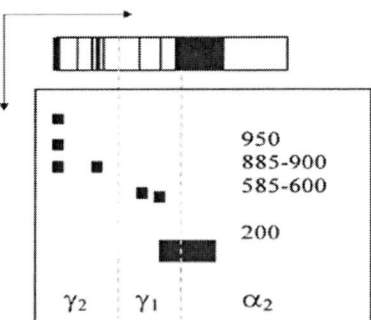

Figure 3.2. Electrophoresis of serumγ- and α-globulins of sterlet in concentration PAGE gradient (3-20%). Vertical pointer show the direction of concentration PAGE gradient electrophoresis, horizontal pointer – of disk-electrophoresis; numerals mean a MM values of proteins in *kDa*. (Andreeva, 2001b).

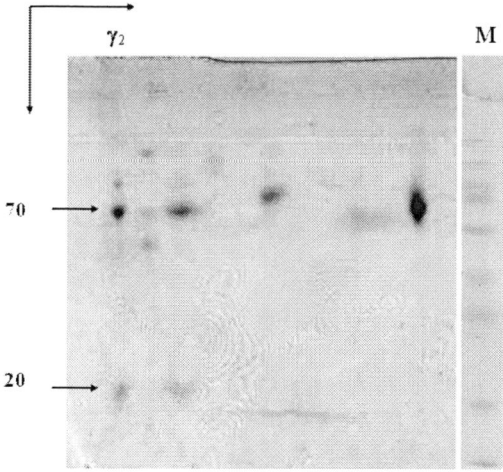

Figure 3.3. SDS-electrophoresis of serumproteins of sterlet (reducing conditions). Vertical arrow show the direction of SDS-electrophoresis, horizontal arrow – of disk-electrophoresis; M – PageRuler™ Prestained Protein Ladder Plus Marker (Fermentas). 20 and 70 – MM values in *kDa* of Ig subunits.

3.3. SERUM ALBUMINS IN THE BLOOD OF *ACIPENSERIFORMES*

Albumin Concentration in Plasma and Blood Serum

Serum albumin is found in the blood of all known *Acipenseriformes*, including the paddlefish (fam. *Polyodontidae*) (Kuzmin, 1996; Andreeva, 1997). The concentration of its total protein in serum is 3-5%. It is practically two times higher than in cartilaginous fishes; the same correlation is also calculated for albumin (Lukyanenko, Khabarov 2005). The dynamic changes of the albumin content in blood of sturgeon fishes occur with changes of water salinity, temperature and fish muscular activity; the serum albumin content in fishes before spawning differs from postspawning ones; the albumin concentration depends on the period of ontogenetic development. All these changes reflects the active participation of albumin in the osmotic regulation of the blood plasma, in the metabolism and in the transport of the important substances, including lipids and carbohydrates (Shulman,1972; Chikhachev, 1982).

The relative content of albumin in the serum of anadromous fishes in the river period of life reachs 30% of the total protein and more, in the sea period of life it is almost two times less (Lukyanenko, Khabarov, 2005). These data allow to assume that an increase in the albumin concentration in the river life period is caused not so by due to osmoregulation, but due to generative function or domination of the plastic function of protein under strengthening of the motion activity of fishes in the fresh water.

Structural Features of Albumins

The identification of the sturgeon albumins on electrophoregrams is not difficult, because their electrophoretic mobility is similar to albumins of mammals (Andreeva, 1986a, 1987c, 2010a). The disk-electrophoresis reveals from 1 to 5 albumin bands in sturgeon fishes serum. The maximum number of albumin phenotypes is 8 and it is found in the sturgeon (Kuzmin, 1996; Kuzmin, Kuzmina, 2005). The large number of allelic variants of sturgeon albumin is the result of the genome polyploidisation, which occurred during the evolution of *Acipenseriformes* (Ludwig et al, 2001; Kuzmin, 2002). The differential manifestation of albumin distinct alleles was revealed in migratory fishes in freshwater and sea water life periods (Kuzmin, 1996). Thus, in Russian sturgeon from the north Caspian Sea population the albumin of phenotype A occurs in 80,5% of individuals; the individuals with the albumin of phenotype AB are not so frequent (19,5%). Meanwhile, in the freshwater period the albumin of phenotype A is found only in 24,7% of the Russian sturgeon population; the single-component albumin of phenotype B is found only in 13,5% of individuals.

Under native conditions different albumin allovariants of sterlet, beluga, starred sturgeon, sturgeon, shovelnose and ship sturgeon were differed by the MM value (from 70 to 79,5 kDa), under the denaturing conditions all albumins had MMapproximately 67 kDa (Andreeva, 1997, 1999). Albumins of all *Acipenseridae* fishes are simple monomeric proteins. Like to human serum albumin the albumins of sterlet and starred sturgeon have one free surface SH-group - 1M to 1M of protein, and adding of 8M urea did not reveal additional "disguised" SH-groups in the albumin (Figure 3.4) (Andreeva, 1987c).

In the presence of 8M urea the major part of blood plasma proteins did not manifest tendency toward aggregation or dissociation, values of their MM remained comparable with the values under nondenaturing conditions (Figure

3.5), but some α_1-globulins in the presence of 8M urea can aggregate (Andreeva, 1997).

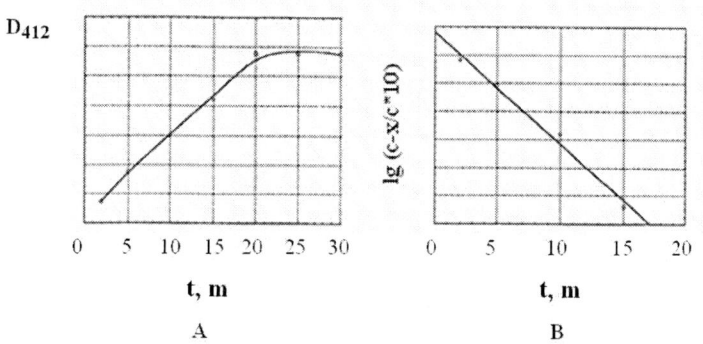

Figure 3.4. Titration of SH-groups of sterlet albumin by dithiobis-2- nitrobenzoic acid (A). Kinetics of modification of SH-groups of sterlet albuminby dithiobis-2-nitrobenzoic acid (B): ordinate axis – lg(c-x/c*10), c – total number of SH-groups titrated in excess of DTNB; x – number of SH-groups, which weretitrated to time moment " t" under even conditions; m - minute. (Andreeva, 1987c).

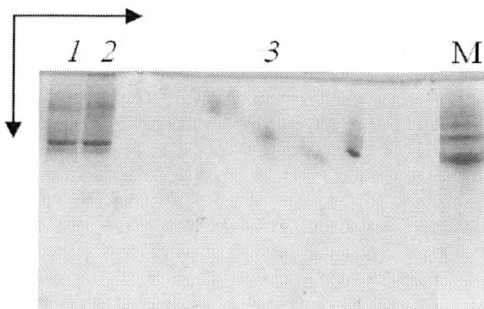

Figure 3.5. Electrophoresis of serum proteins of sterlet in PAGE with 8M urea: 1, 2, 3 - sterlet serum, M – HSA. Vertical arrow shows the direction of electrophoresis PAGE with 8M urea, horizontal arrow – of disk-electrophoresis.

The similarity of albumins of sturgeon fishes with the mammalian ones is revealed at their binding with the albuminspecific dye - bromcresol purple.

The binding was accompanied by the specific shift of λ_{max} from 590 to 603 nm (Figure 3.6) (Andreeva, 1985).

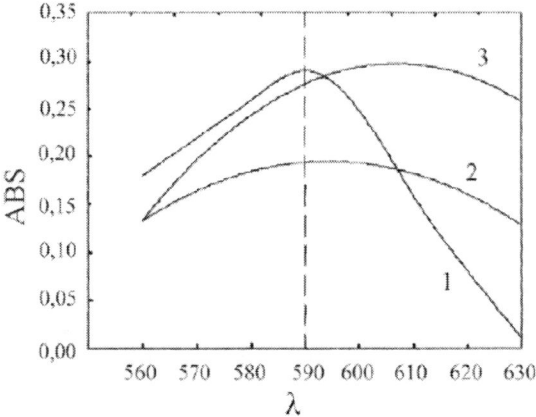

Figure 3.6. Absorption spectrum of bromocresol purple BCP (1) and complexes BCP-albumin of bream (2) and stellate sturgeon (3). λ - wavelength, nm. (Andreeva, 1985).

The similarity of albumins of sturgeon fishes and mammals is revealed in the parameters of albumins salting out by ammonium sulfate and by sodium sulfite, in the solubility in ethanol and in the mixture of ethanol and trichloracetic acid, in the precipitation by rivanol (Chikhachev, 1982). Like mammalian albumins the sturgeon fishes albumins did not contain carbohydrate in the structure of molecule (Andreeva, 1999).The analysis of the amino-acid composition of sturgeon fishes albumins confirms its similarity to albumins of other vertebrates. The analysis of amino-acid sequence reveals the identity of the majority of links of NH_2-terminal fragment of the white sturgeon albumin and of some vertebrate species (citation from Chikhachev, 1982).

Thus, *Acipenseriformes* albumins are the simple monomeric proteins, which have one surface free SH- group. Sturgeon albumins differ from albumin-like proteins of sharks and rays. They are amazingly similar to the albumins of mammals and contemporary representatives of lungfishes, originated from those *Rhipidistii*, which gave the beginning of first tetrapods. This is an example of the convergent evolution of albumins in the posterity of *Rhipidistii* and *Palaeonisci*.

Chapter 4

ORGANIZATION OF *TELEOSTEI* BLOOD PLASMA PROTEINS

4.1. ENVIRONMENT AND THE COMPOSITION OF THE EXTRACELLULAR FLUIDS IN THE ORGANISM OF BONY FISHES

Real bony fishes originated, probably, from the different groups of the ancient bony ganoids – amie fishes *Amiiformes* and *Lepisosteiformes*, that supposedly originated from *Palaeonisci*. Among *PiscesTeleostei* are distinguished by high fertility and adaptability, variety of ecological niches. They are represented by fresh water, brackish water and sea species, nonmigratoty and migratory forms. Some fresh water fishes are capable to inhabit waters with the salinity from 0,5 to 30‰.

The urea content in the blood of the majority of the bony fishes is 100 times less, and TMAO - 10 times less in comparison with the sharks; and the salts content varies substantially (Shilov, 1985; Martemyanov, 2001).

4.2. SERUM GLOBULINS

All charactеric for vertebrates globulins - transferrin, hemopexin, immunoglobulins and haptoglobin are present in the blood of bony fishes.

Transferrins

There is usually only one or a few working transferrin loci in the fish genome. The transferrin locus is variable, the number of its alleles changes in the range from 2 to 13, more frequently – 3-4; the alleles are by codominant. The number of transferrin allovariants varies correspondingly. The heterogeneity of more than three components (bands) in the electrophoresis is caused by sialic acids and conformational changes of transferrin (Kirpichnikov, 1987).

It was considered that the polyploid bony fish species (carp, two crucian species, the barbel *Barbus barbus*, all *Catostomidae* and *Salmonidae* fishes) have also only one transferrin locus (Kirpichnikov, 1987). Later two transferrin loci were revealed in *Salmonidae* fishes - *Salmo salar* and *Salmo trutta* (Kvingedal et al., 1993; Kvingedal, 1994; NCBI GenBank: ACC55226).

The transferrin gene«**tfa**» of zebrafish *Danio rerio* (*Brachydanio rerio*) encodes transferrin-a, which consists of 673 amino-acid residues (MM 73282 Da) (UniProtKB/TrEMBL: B8JL43 (B8JL43_DANRE); NCBI, tfa transferrin-a [*Danio rerio*]) (Figure 4.1. 4.2); the larva transferrin (transferrin-a)consists of 520 amino-acidresidues(UniProtKB/TrEMBL:Q567C8). Thetransferrin geneof nile tilapia(Tilapia nilotica)*Oreochromis niloticus*encodesthe protein, whichconsists of 694amino-acid residues (UniProtKB/TrEMBL: A2SUL9).The transferrin of rainbow trout*Oncorhynchus mykiss* (*Salmo gairdneri*) consists of 691amino-acid residues (UniProtKB/TrEMBL: Q9PT13);of grass carp*Ctenopharyngodon idella* (*Leuciscus idella*) - of 615 amino-acid residues (UniProtKB/TrEMBL:Q6TXT2); of silver crucian carp*Carassius gibelio*– of 669amino-acid residues (UniProtKB/TrEMBL:Q90WL8 (Q90WL8_9TELE)), of 671 (variant D)(UniProtKB/TrEMBL: Q8JHD4 (Q8JHD4_9TELE)), of 661 (variant C) (UniProtKB/TrEMBL: Q8UVE8 (Q8UVE8_9TELE)), of 666 amino-acid residues(transferrin variant E, 73462 Da) (UniProtKB/TrEMBL: Q8JHD3 (Q8JHD3_9TELE)); of crucian carp*Carassius cuvieri* - of671amino-acid residues (transferrin variant C, MM 73990 Da) (Figure 4.3) (UniProtKB/TrEMBL: Q7T1G7 (Q7T1G7_CARCW));of medaka fish*Oryzias latipes* (Japanese ricefish) – of 690 amino-acid residues (UniProtKB/TrEMBL: A8MN21), of common carp*Cyprinus carpio* – of 669 amino-acid residues (variant A) (UniProtKB/TrEMBL:Q8UVE7(Q8UVE7_CYPCA)),of silver carp*Hypophthalmichthys molitrix* (Leuciscus molitrix) – 674 amino-acid residues (UniProtKB/TrEMBL: E2IU44), of bighead carp*Hypophthalmichthys nobilis* (Aristichthys nobilis) – of 671amino-acid residues

(UniProtKB/TrEMBL: D9MX86), of channel catfish*Ictalurus punctatus* – 679 amino-acid residues (UniProtKB/TrEMBL: C9W3Q0); of brown trout*Salmo trutta*(UniProtKB/TrEMBL: Q9PU66), of biwa salmon*Oncorhynchus masou rhodurus* (UniProtKB/TrEMBL: Q789D8), of sockeye salmon*Oncorhynchus nerka* (UniProtKB/TrEMBL: Q9PU70) and cherry salmon or masu salmon*Oncorhynchus masou* (UniProtKB/TrEMBL: Q9PRH5691) – 691 amino-acid residues etc.

Transferrins of bony fishes are glycoproteins with MM approximately 70 kDa (Valenta et al, 1976) and are organized as monomeric proteins. The spectral characteristics of fish and mammalian transferrins differ: λ_{max} for bream transferrin is 410 nm, for human one it is 465 nm (Andreeva, 1987a, 1997).

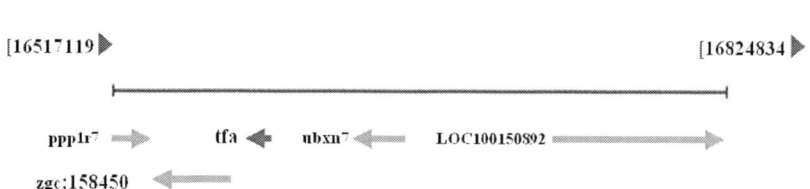

Figure 4.1. Location of zebrafish transferrin-a gene on 2th chromosome. RefSeq status PROVISIONAL (NCBI Gene, tfa transferrin-a [*Danio rerio*].

Immunoglobulins

The majority of *Osteichthyes* fish species, including *Teleostei*, has IgM (Marchalonis, Schonfeld, 1970; Kobayashi et al., 1982; Mochida et al., 1994; Nakamura et al., 2006; Grove et al., 2006). An addition to IgM the immunoglobulins IgD and IgZ/T (orthologists)are detected in *Teleostei* (Hansen et al., 2005; Hsu et al., 2006). The MM value of immunoglobulins in the Atlantic halibut *Hippoglossus hippoglossus* is 780 kDa (Grove et al, 2006), in the live-bearing fish *Neoditrema ransonneti* - 820 kDa (Nakamura et al, 2006), in the bream *Abramis brana* L. - 950, 700 and 510 kDa (Andreeva, 2001a). The number of amino acids in the Ig composition of the rainbow trout *Oncorhynchus mykiss* and of many other fish species varies from 551 to 600 (Hansen et al, 2005). By the example of Atlantic salmon*Salmo salar* it is shown, that the duplication of immunoglobulin loci*IGH* (*IGH-A* and *IGH-B*) leads to the considerable increase in the antibodies variety in comparison with other vertebrates (Yasuike et al, 2010).

```
         10         20         30         40         50         60
 MKVLLISLLG CLVVALPSAS AQKKVKWCVT TQNEQSKCRH LATKAADIEC HLQPTVIDCM
         70         80         90        100        110        120
 RSIAAGGTDI VTVDGANVFT GGLNNYLLRP IIAEKKKECC YAVAAVKAGS GFNINELKGK
        130        140        150        160        170        180
 SSCHSCYQRS GGWNTPIGKL IATNKITWEG PNEMPVERAV SEFFSSSCVP GVSKPKYPNL
        190        200        210        220        230        240
 CKACQGDCSC SHNEKYFGDD GAFQCLKNDN GQVAFVCHHA IPESERQNYE LLCMDGSRKS
        250        260        270        280        290        300
 VEDYKTCNFA REPARTVIAR TDTDLQYVYD VLKQIPASDL FSSQAFGGKD LIFSDSATEL
        310        320        330        340        350        360
 MLLPKRTDSL LYLKEEYYEA MQAFKDGNPS APTSQTKLAM CTIGHAEKNK CDSLDHVKKS
        370        380        390        400        410        420
 CILEASVDDC IEKIKRKEAD FLAVDGGQVY IAGKCGLVPV MAEQSNSQSC SSGSGGTASY
        430        440        450        460        470        480
 YAVAVVRKGS GLTWNNLEGK KSCHTGLGRS AGWKIPESAI CGEKDKCTLD KFFSEGCAPG
        490        500        510        520        530        540
 ADPTSNMCKL CKGSGKAVGD ESKCKPSAEE QYYGYDGAFK FRCLAEKAGD VAFIKHTVVG
        550        560        570        580        590        600
 DYTDGKGKDW AKDLKSEDFE LICPNTPDTT MKYTDFEKCN LAQVPVHAVI TREDARSAVV
        610        620        630        640        650        660
 SFLSDIQSKN NDLFTSKDGK NLLFTDGTKC LQEIKGSVDD FLTKKYIDMI ERTYKTSQNV
        670
 PDLVKACTFG NCISS
```

Figure 4.2. Amino acid sequence of transferrin-a (complete) of zebrafish *Danio rerio* (*Brachydanio rerio*). Sequence length 673 AA, Mass 73,282 Da. UniProtKB/TrEMBL: B8JL43 (B8JL43_DANRE).

```
         10         20         30         40         50         60
MNIPLISFLA CLVVALPSES AQKVKWCVKS QHELKKCQYL ATKSPELECH LKSSVTECMI
         70         80         90        100        110        120
SIKTGEADAI TVDGEHVYQA GLINYDLRPI IAEDCKAVCS YAVALVKRDT DFSINDLKGK
        130        140        150        160        170        180
TSCHSCYQSP GGWDIPIGRL VKEHKIPWDG IDDMPLEKAV SQFFSSSCIP GISKAVYANL
        190        200        210        220        230        240
CQGCQGDCSC SDSEKYSGDG GAFQCLKSGH GQVAFMCHDG VPSSERQNYQ LLCMDGSRKS
        250        260        270        280        290        300
VEEYKDCYFL KEPCHAVISR KDADSEQIYK VLQQIPASDL FSSAAFGGKD LMFSDPPTEL
        310        320        330        340        350        360
TELPKSMDSF LYLREDYYEA MRALRDGNPK DPPQDGKIQW CIISHAEQQK CDSLQIPRME
        370        380        390        400        410        420
CRRTSSVEEC IQKIMRKEAD ALTVDGGQVY IAGKCGLVPV MVEQSDQQSC PDGGEASSYY
        430        440        450        460        470        480
VVAVVRKASG VTWNTLKGKK SCHTGLNRNA GWKVPDSAIC GKTPGCTLYN FFSKGCAPGA
        490        500        510        520        530        540
DPKSNMCELC KGSGKAVGDE SKCKASSEEK YYGYDGAFRC LAEKTGEVAF IKHTIVGDYT
        550        560        570        580        590        600
DGKGPEWAKD LKSEDYELIC PESPDTTVKH TEFVRCNLAK VPAHAVITRE DARKDVVKVL
        610        620        630        640        650        660
KEAQANSDKL FKSEGERNLL FSDSTKCLQE ITQPLKEFLT QEYIDMIEKT YTTGQGKPDL
        670
VKACTIKMCF G
```

Figure 4.3. Amino acid sequence of transferrin (variant C) (complete) of *Carassius curveri* (Crucian carp). Sequence length 671 AA, Mass 73,990 Da. UniProtKB/TrEMBL: Q7T1G7 (Q7T1G7_CARCW).

The high-molecular immunoglobulins of *Teleostei* consist of monomeric moleculeswithMM approximately 180 kDa, which aggregate into the oligomeric complexes and the high-molecular aggregates at the expense of the noncovalent bonds. The IgM of Nile tilapia *Oreochromis niloticus* is represented by the tetrameric protein with MM about 760 kDa, which does not have the J-chain, while in cartilaginous fishes and mammals it is a pentamer (Mochida et al, 1994). The bream serum immunoglobulins with MM approximately 180 kDa aggregate into thecomplexes with MM about 510, 700 and 950 without the involvement of covalent bonds (Andreeva, 2001a).

In electrophoresis the fish serum immunoglobulins are located in the zone of the mobility of gamma globulins - practically at the beginning of the electrophoregram. The bream gamma globulins are eluated in two peaks after the fractionating on the column with Sephadex G-200(Figure 4.4).

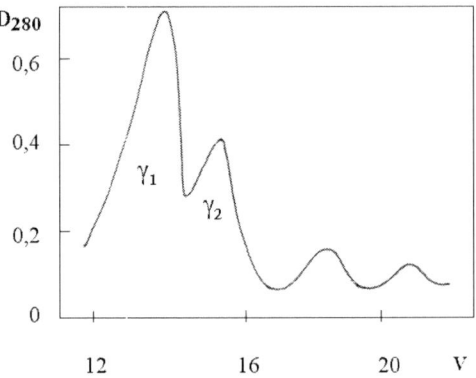

Figure 4.4. Elution profile of bream γ-globulins (γ_1 and γ_2) from column with Sephadex G-200, 0.1M tris-HCl, pH 8.9. γ-globulins were fourfold reprecipitate in ethanol. V – volume, milliliter.

The first peak coincides with the free volume of column and containes gamma-1-globulins with MM about 320, 280 and 180 kDa (Figure 4.5).

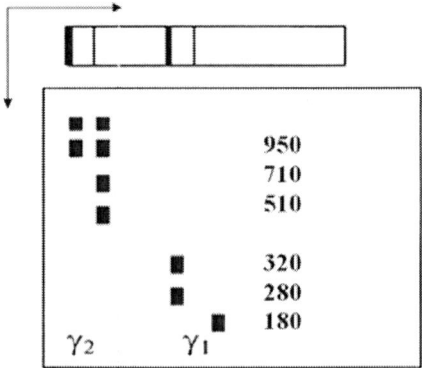

Figure 4.5. Scheme of electrophoresis of bream serumγ-globulins in concentration PAGE gradient (3-20%). Vertical arrow show the direction of concentration PAGE gradient electrophoresis, horizontal arrow – of disk-electrophoresis; numerals mean a MM values of proteins in *kDa*.(Andreeva, 2001b).

The second peak containes the proteins with MM approximately 180 kDa. They are located in the zone ofgamma-2-globulins mobility - practically at the start of the electrophoregram. These proteins are represented by the aggregates with MM approximately 510, 700, 950 and above 1000 kDa in the concentration gradient of PAGE (Figure 4.5) (Andreeva, 2001a).

The proteins with the Y-similar structure are discovered in the gamma-2-globulins subfraction. These proteins are decomposed into 5-6 chains with MM from 40 to 90 kDa and 1-3 chains with MM approximately 20 kDa in the 2D-SDS-electrophoresis (in presence of reducing agent) (Figure 4.6) (Andreeva, 2001a).

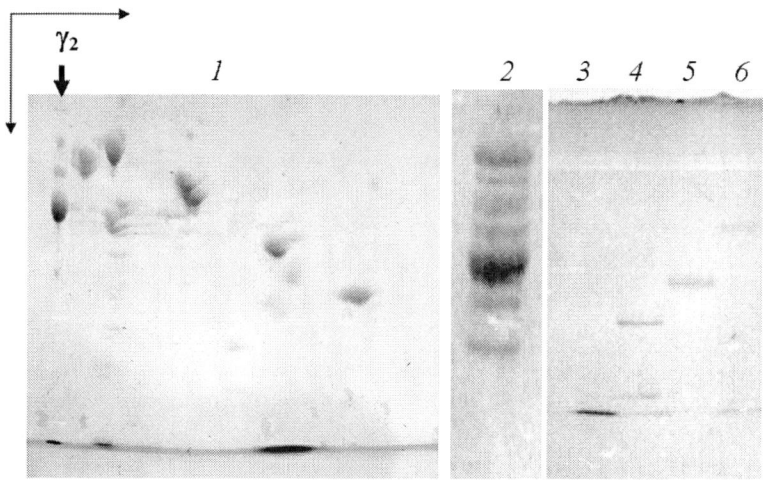

Figure 4.6. 2D-SDS-electrophoresis of bream serum proteins (*1, 2*) (reducing conditions). Vertical arrow show the direction of SDS-electrophoresis, horizontal arrow – of disk-electrophoresis; *3-6* – MM markers ribonuclease; troponins T, I, C; ovalbumin and BSA respectively. Little vertical pointer indicate to chains of immunoglobulins from γ_2-globulin fraction. (Andreeva, 2008).

The number of fish Ig genes can reach 100 (Hsu et al, 2006). IGL and IGH genes are organized differently. Several variants of L-chains have been revealed in the structure of the channel catfish*Ictalurus punctatus*. The channel catfish IGL genes belong to the kappa-type (Ghaffari, Lobb, 1993, 1997; NCBI GenBank: AAA16654). L-chains are divided into two classes - F and G. The channel catfish G-type IGL loci are organized according to the cluster type (Figure 4.7) (Ghaffari, Lobb, 1993, 1997).

Figure 4.7. Scheme of the structure of the loci of IGL of channel catfish*Ictalurus punctatus*(according to Ghaffari, Lobb, 1993).

The six variants of IG L-chains with MM from 25 to 28,5 kDa are revealed in the Atlantic halibut *Hippoglossus hippoglossus* (Grove et al, 2006); three L-chain isotypes - in Atlantic salmon*Salmo salar* (Solem, Jorgensen, 2002); one L-chain isotype – in the Fugu fish (Sphaeroides, the family *Tetraodontidae*) (Saha et al, 2004). There are two types of L-chains in the rainbow trout *Oncorhynchus mykiss* Ig, one of them is high by homologous to the G- type IGL of the channel catfish*Ictalurus punctatus*, and the corresponding loci is also organized according to the cluster type (Daggfeldt et al, 1993)

The IG heavy chains of the bony fish are organized not according to the cluster, but according to the segmental type (Hsu et al, 2006). The reason for this is possibly the plural transposon families, which induce genetic reconstructions (Volff, 2005).The µ-similar heavy chains with MM approximately 70 kDa and H-chains with MM approximately 38 and 52 kDa (Litman et al, 1971) and 76 kDaare described (Grove et al, 2006) and the chimerical analog of channel catfish IgD heavy chain is discovered (Wilson et al, 1997); the heavy chain (IgT) with MM approximately 63 kDa for the membrane-bound form and 58 kDa for secreting one are described in the rainbow trout (Hansen et al, 2005).

The variety of chains of serum Ig from γ_2-globulim subfraction of plasma and organism tissue fluids (from the brain, the white muscles and the peritoneal fluid) is revealed in the hogfish or black scorpionfish *Scorpaena porcus* L. Thus, 10 different subunits with MM from 25 to 200 kDa have been revealed in the structure of immunoglobulins from peritoneal fluid in SDS-PAGE. The MALDI-TOF-analysis of four subunits with MM approximately 65, 70, 80 and 90 kDa demonstrated practically the complete identity of their mass-spectra (Andreeva, Dmitrieva, 2011) (Figure 4.8).

The search for homologues for scorpaena heavy chains in NCBI database among the proteins of all organisms with the help of the program Mascot (www.matrixscience.com) revealed the homologue only for the subunit with MM approximately 70 kDa

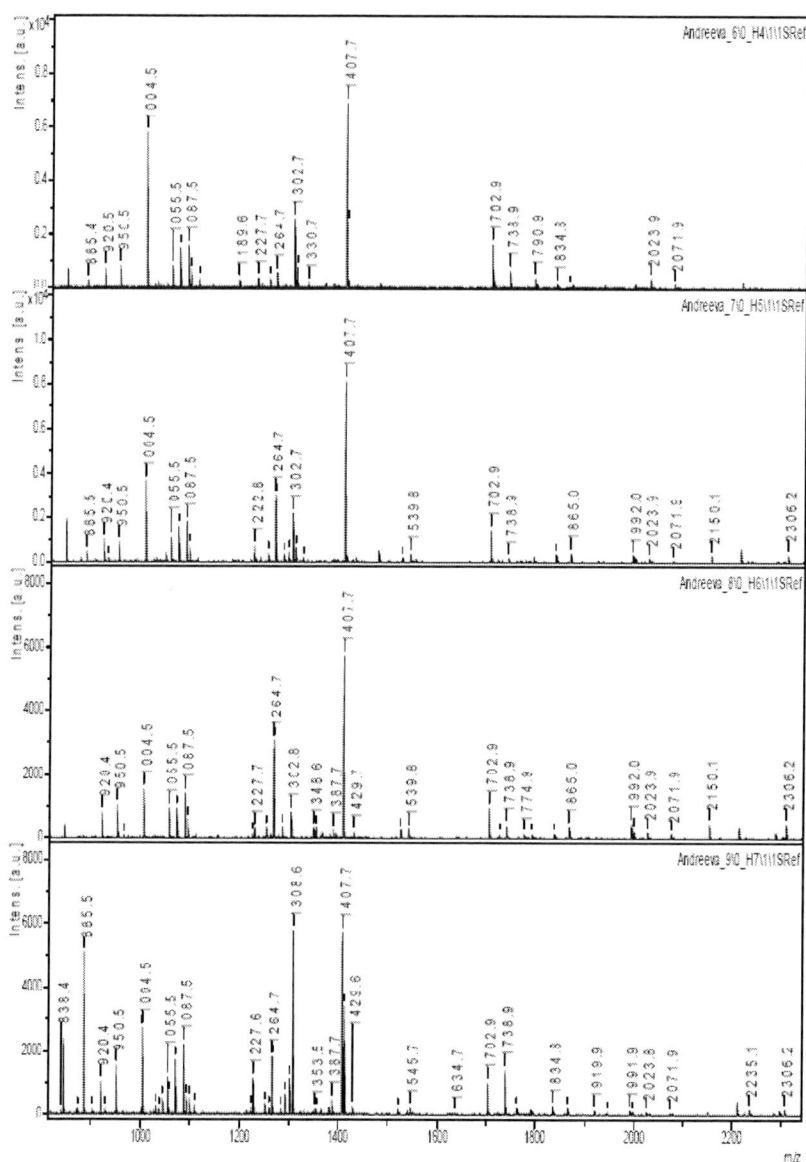

Figure 4.8. Mass-spectra MS of scorpaena Ig H-chains. The abscissa axis is the values of tryptic cleavage fragments MM, the ordinate axis is the signal intensity. (Andreeva, Dmitrieva, 2011).

The gomolog is H-chain of chinese perch *Siniperca chuatsi* immunoglobulin with MM 53294 Da (NCBI GenBank: AAQ14846). The fragment ***NGAALTDSIQYPPVQK*** (MM 1702.9 Da) in its composition has been discovered in scorpaena Ig (Figure 4.9).

```
  1 EDTAVYYCAR AVLGVYGFDY WGKGTMVTVS SATSTGPTVF PLMQCGSGTG

 51 DMVTLGCLAT GFTPSSLTYA WSKNGAALTD SIQYPPVQKG DVYTGVSQIR

101 VRRQDWDARE SFRCAVTHPA GNGKADFMKP KVTYVPPTEL KVLASSGEEQ

151 EASFSCFARD FSPKDYEIKW LKNEAEIPNK IYEIKMPLGE RQDKNGTTLY

201 SAASFLTVPA SEWTVDTKFT CEFEGKGENG ATFMNSSVTY KHTTPGNCEV

251 DVDIKITGPT LADMFLNREG TIVCQVKVNE PYVGRILWED EKGNEMAGAS

301 KTFNDEGTFS LPLEITYDEW SKGIKRYCVV EHENLIEPLK ELYERSFGGQ

351 TQRPAVFMLP PVEHTRKETV TLTCYVKDFF PQEVLVTWLV DDEEADSKYK

401 FYTTNPVESN GSYFAYGQLS LSLEQWKKND VVYSCVVHHQ SLVNTTNAIV

451 RSIGHRTFEN TNLVNLNMNI PESCKAQ
```

gi|33340500|gb|AAQ14846.1|AF320776_1 immunoglobulin heavy chain [Siniperca chuatsi]

Figure 4.9. The amino acid sequence of the Ig heavy chain of the chinese perch *Siniperca chuatsi*. The sequence fragment labeled by italic type *NGAALTD SIQYPPVQK* (74-89 position) is gomolog to scorpaena Ig H-chain.

The candidate proteins are the definitely reliable homologues ($p < 0.05$) only if they have the parameter of the reliable *score* > 83. The candidate the heavy chain Ig of chinese perch *Siniperca chuatsi* revealed in the result of the search had a parameter *score* 57; therefore it was examined as the possible homologue of the scorpaena protein. This example shows that the plurality of H-chains can be formed due to the homologous variants of one chain within the one class of Ig (Andreeva, Dmitrieva, 2011).

Haptoglobins

Serum haptoglobins bind the hemoglobin, which appeares in the blood plasma after the intravascular hemolysis of erythrocytes. For the first time the haptoglobin appears among Vertebrata in the group of bony fishes as the specialized protein. It is discovered and described in the sea perches of the genus*Sebastes* (Nefedov, 1969), in goldfish *Carassius auratus gibelio* (Poljakovskij et al., 1973), in European carp *Cyprinus carpio* (Chutaeva et al., 1975), in the European and Far Eastern *Cyprinidae* (Karnauhov, 1987; Andreeva, 2001b), in grass carp (Leuciscus idella) *Ctenopharyngodon idella*(fragment, 59 amino-acid residues) (UniProtKB/TrEMBL: Q6SXN9), in rainbow trout *Oncorhynchus mykiss*(fragment, 108 amino-acid residues)(UniProtKB/TrEMBL: Q9I8H1(Q9I8H1_ONCMY)) (Figure 4.10);rainbow trout haptoglobin-1 (fragment, 95 amino-acid residues)(UniProtKB/TrEMBL: Q9DFG1 (Q9DFG1_ONCMY)) and haptoglobin-2 (fragment, 77 amino-acid residues)(UniProtKB/TrEMBL: Q9DFG0 (Q9DFG0_ONCMY)) are described.

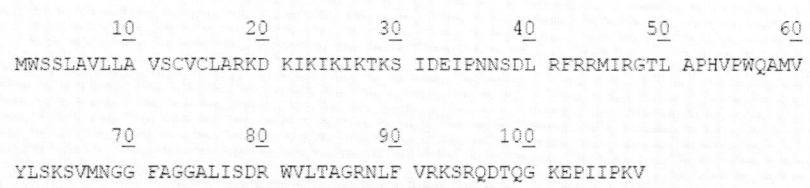

Figure 4.10. Amino acid sequence of haptoglobin (fragment, Length 108 AA, Mass 12,095 Da) from rainbow trout *Oncorhynchus mykiss*haptoglobin.UniProtKB/ TrEMBL: Q9I8H1 (Q9I8H1_ONCMY).

Haptoglobin gene *Danio rerio* is located on 8th chromosome (Figure 4.11).

The structure of the fish haptoglobin has its specific features. It is distinguished in different fish species and does not coincide with the haptoglobin structure of other vertebrates (Nefedov, 1969). The haptoglobin of the Zherekh *Aspius aspius* has more than one hemoglobin-binding site. One type of the haptoglobin molecules is discovered in four species of the *Catastomidae* family and its different electrophoretic mobility is caused by the different ratio of haptoglobin and hemoglobin in the haptoglobin-hemoglobin complex: the relation in some complexes haptoglobin- hemoglobin is 1:1 and in others 1:2 (Koehn, 1966).

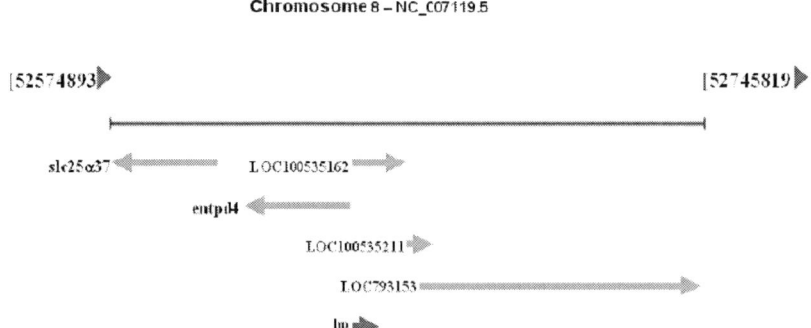

Figure 4.11. Model of location of haptoglobin gene of Danio rerio on 8th chromosome.NCBI Gene, hp haptoglobin [*Danio rerio*].

The hemoglobin-binding proteins are revealed in the alpha-2-globulin serum fraction of the bream and roach. The alpha-2-globulin fraction proteins from the laky serum are constantly stained by bezidine reagent. And in 25% of nonhemolized serum tests this fraction are not stained by bezidine reagent. The incubation of such sera with the excess of native hemoglobin leads to the restoration of the peroxidase activity of this fraction, that is caused, probably, by the presence of the hemoglobin-binding protein in it. In 75% of nonhemolized sera alpha-2-globulins are stained by bezidine, that is probably caused by the presence of binding hemoglobin in this fraction (Andreeva, 2001b).

The detection of hemoglobin binding by alpha-2-globulins proves the of the hemoglobin-binding protein - haptoglobin presence in this fraction. In 2D-PAGE with 8M ureaof bream serumstarting proteinsand proteins with MM approximately 270 kDa located on the track of alpha-2-globulins; in 2D-SDS-PAGE seven subunits with MM from 10 to 46 kDa and 20 subunits with MM from 50 to 100 kDa located on this track (Figure 4.12).

The human serum contains the haptoglobin, which consists of the subunits with MM approximately 19,8 (and/or 9) and 42,6 kDa (Chapter 1). Probably low-molecular subunits of the bream haptoglobin with MM, close to these values, enter into the composition of the haptoglobin molecule with MM approximately 270 kDa, similar to human haptoglobin of genetic type Hp 2-1.So the fish haptoglobins are represented by different genetic variants with different MM values. They are monomeric molecules with the subunit structure.

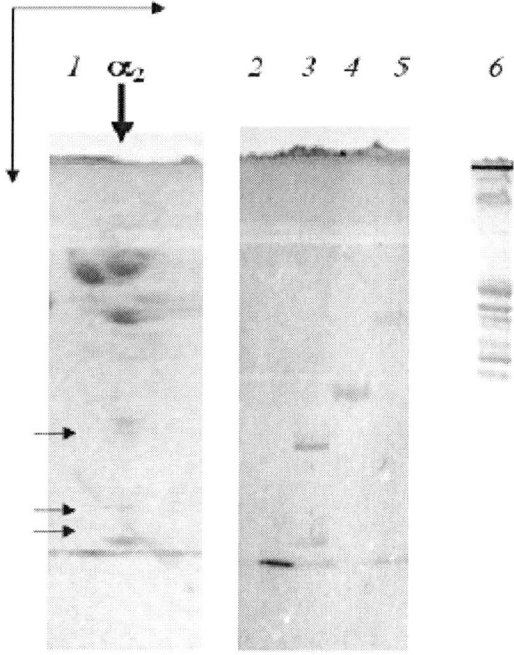

Figure 4.12. 2D-SDS-electrophoresis of bream serumα_2-globulins (*1*). Vertical arrow show the direction of SDS-electrophoresis, horizontal arrow – of disk-electrophoresis; *2-5* – MM markers ribonuclease; troponins T, I, C; ovalbumin and BSA respectively; *6* – bream blood serum. Little vertical pointer indicate to track of α_2-globulins and haptoglobin in its composition. Little horizontal pointers indicate to subunits with MM about 16, 25 and 45 kDa. (Andreeva, 2001a).

Hemopexin

Hemopexins are detected in Zebrafish *Danio rerio* (UniProtKB/Swiss-Prot: Q6PHG2 (HEMO_DANRE)) (Figure 4.13), *Hypomesus transpacificus* (UniProtKB/TrEMBL: C3UVG6 (C3UVG6_9TELE)). Hemopexin-like protein is detected in Rainbow trout *Oncorhynchus mykiss* (Salmo gairdneri) (UniProtKB/TrEMBL: P79825 (P79825_ONCMY)) (Figure 4.14) respectively.

```
          10         20         30         40         50         60
    MRLIQALSLC LALSLSLAAP PQHKEDHSHK GKPGGEGHKH ELHHGAQLDR CKGIEFDAVA
          70         80         90        100        110        120
    VNEEGVPYFF KGDHLFKGFH GKAELSNKTF PELDDHHHLG HVDAAFRMHS EDSPDHHDHQ
         130        140        150        160        170        180
    FFFLDNMVFS YFKHKLEKDY PKLISAVFPG IPDHLDAAVE CPKPDCPNDT VIFFKGDEIY
         190        200        210        220        230        240
    HFNMHTKKVD EKEFKSMPNC TGAFRYMGHY YCFHGHQFSK FDPMTGEVHG KYPKEARDYF
         250        260        270        280        290        300
    MRCPHFGSKT TDDHIEREQC SRVHLDAITS DDAGNIYAFR GHHFLSITGD KFHSDTIESE
         310        320        330        340        350        360
    FKELHSEVDS VFSYDGHFYM IKDNDVFVYK VGKPHTHLEG YPKPLKDVLG IEGPVDAAFV
         370        380        390        400        410        420
    CEDHHVVHII KGQSIYDVDL KATPRKLVKE GTITQFKRID AAMCGPKGVT VVIGNHFYNY
         430        440
    DSVQVMLMAK IMPEQQKVSQ QLFGCDH
```

Figure 4.13. Amino acid sequence of Hemopexin (complete) from Zebrafish *Danio rerio*. Length 447 AA, Mass 51,027 Da. UniProtKB/Swiss-Prot: Q6PHG2 (HEMO_DANRE).

```
          10         20         30         40         50         60
    TMKPLSQTLC LCLVLALSHA HHHAGHQGGE DEGHEGHDHG HHEGLLLDRC QGIEMDAVAV
          70         80         90        100        110        120
    TEEGIPYFFK GGHVFKGFHG KAELSNESFA ELDDHHHLGH VDAAFLMHFP DKPTEHDHIF
         130        140        150        160        170        180
    FMLDTKVFSY YKHQLETGFP KDISEVFPGI PDHLDAAVVC PAPDCEEDAV IFFKGDEIYH
         190        200        210        220        230        240
    YNVKTKKVEE KKFEGMPNCT SAFRFMEHYY CFHGHQFSKF DPKTGEVHGR YPKEARDYFM
         250        260        270        280        290        300
    KCSKFGDTTD HIERERCSRV HLDAITSDDA GNIYAFRGHH FLEQDAGNDT WAADTIESDF
         310        320        330        340        350        360
    KELHSEVDAT FSYENHLYMV KDDKVYIYKV GDSHTHLDGS PKPLKEVLGV EGPIDAAFVC
         370        380        390        400        410        420
    QDHHIAHVIK GQTVYDVLLK ASPPVPVKEG SFTLFNKVLA AMCGPEGVKL FKGNHYFHFQ
         430        440
    SVKVMLMAKA TPEEHKTALE LFGCDH
```

Figure 4.14. Amino acid sequence of hemopexin-like protein (fragment) from Rainbow trout *Oncorhynchus mykiss*. Length 446 AA, Mass 50,455 Da. UniProtKB/TrEMBL: P79825 (P79825_ONCMY).

Gene hpx of *Danio rerio* is locater at chromosome 9 (Figure 4.15).

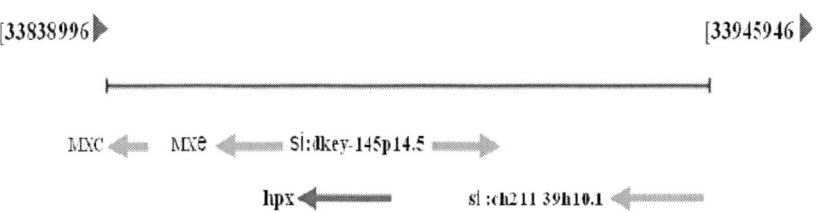

Figure 4.15. Location of zebrafish hemopexin gene on 9th chromosome. RefSeq status PROVISIONAL (NCBI Gene, hpx hemopexin [*Danio rerio*].

4.3. SERUM ALBUMIN CONTENT IN THE BLOOD OF *TELEOSTEI*

The albumin concentration in the bony fish blood plasma can significantly vary, especially in the group of the highest *Teleostei*. The albumin concentration in the blood of marine bony fish blood is lower than in fresh water one, as in the case of *Acipenseriformes* (Chapter 3) (Shulman, 1972). Variations of the albumin concentration depend on the degree of the natural mobility of the species, on the character and duration of loads, season, stage of the maturity of the gonads and other factors (Kirsipuu, Laugaste, 1979; Bede, 1959). In mobile fish species (tunny) the relative content of albumins can reach 60% of the total protein, and in low-mobile fish species (goby) it does not exceed 5% (Kejvanfar, 1962).

The seasonal dynamics of the albumin level in the blood has a feedback with the protein-synthesizing activity of the hepatocytes (Kirsipuu, Laugaste, 1979).The generative dynamics of albumin also depends on the activity of the hepatocytes. The albumin-synthesizing activity of hepatocytes (hepatocyte culture), determined by the level of albumin mRNA, depends on the dose and the time of the action of estradiol (Flouriot et al, 1998). The intensive expense of albumins for the plastic needs occurs in the active period of life, both their synthesis and expense practically end in winter, and in spring a certain catabolism of albumins begins (Mouridsen, Wallevik, of 1968; Mouridsen,

1969), as a result the albumin number in the serum decreases and this is the signal for the beginning of the active synthesis of this protein by the hepatocytes (Kirsipuu, Laugaste, 1979).

4.4. THE PROBLEM OF *TELEOSTEI* SERUM ALBUMIN IDENTIFICATION

The albumin identification in bony fishes is hindered in a number of cases. It is caused by the atypical arrangement of the mobile plasma fractions on electrophoregrams. The electrophoretic mobility of these fractions does not correspond to the mobility of mammalian albumins. Low-molecular fractions are represented often in the electrophoresis by plural bands or "spots" with the illegible outlines, "floating" to other fractions (Andreeva, 1999, 2010a, b). This leads to the confusion in the terminology: some authors use the term "albumin-similar protein", others - "albumin", the third authors use the term analbuminemia (Sulya et al, 1961; De Smet et al, 1998).

The opinion exists that analbuminemia is characteristic for the less specialized groups of bony fishes, which preserved the ancient features of organization, and among the highest fishes the analbuminemia occurs rarely. During the detection of albumins in the first group of fishes, their concentration in the blood is considerably lower than in the individual representatives of the second group (Sulya et al, 1961). However, this viewpoint is not fconfirmed and the albumins are discovered in the less specialized groups of bony fishes with the ancient features of organization.

Family Salmonidae

The albumins of Atlantic salmon*Salmo salar*, Chinook salmon *Oncorhynchus tshawytscha*, brook trout *Salmo trutta*, Rainbow trout *Oncorhynchusmykiss*and others are identified and described (Figure 4.16, 4.17, 4.18).

Albumin concentrations in their plasma are rather high - about 15 mg/ml (Byrnes, Gannon, 1990; Metcalf et al, 1998ab; Xu, Ding, 2005).

```
         10          20          30          40          50          60
MQWLSVCSLL  VLLSVLSRSQ  AQNQICTIFT  EAKEDGFKSL  ILVGLAQNLF  DSTLGDLVPL
         70          80          90         100         110         120
IAEAIAMGVK  CCSDTPPEDC  ERDVADLFQS  AVCSSETLVE  KNDLKMCCEK  TAAERTHCFV
        130         140         150         160         170         180
DHKAKIPRDL  SLKAELPAAD  QCEDFKKDHK  AFVGRFIFKF  SKSNPMLPPH  VVLAIAKGYG
        190         200         210         220         230         240
EVLTTCCGEA  EAQTCFDTKK  ATFQHAVMKR  VAELRSLCIV  HKKYGDRVVK  AKKLVQYSQK
        250         260         270         280         290         300
MPQASFQEMG  GMVDKIVATV  APCCSGDMVT  CMKERKTLVD  EVCADESVLS  RAAGLSACCK
        310         320         330         340         350         360
EDAVHRGSCV  EAMKPDPKPD  GLSEHYDIHA  DIAAVCQTFT  KTPDVAMGKL  VYEISVRHPE
        370         380         390         400         410         420
SSQQVILRFA  KEAEQALLQC  CDMEDHAECV  KTALAGSDID  KKITDETDYY  KKMCAAEAAV
        430         440         450         460         470         480
SDDSFEKSMM  VYYTRIMPQA  SFDQLHMVSE  TVHDVLHACC  KDEQGHFVLF  CAEEKLTIAI
        490         500         510         520         530         540
DATCDDYDPS  SINPHIAHCC  NQSYSMRRHC  ILAIQPDTEF  TPPELDASSF  HMGPELCTKD
        550         560         570         580         590         600
SKDLLLSGKK  LLYGVVRHKT  TITEDHLKTI  STKYHTMKEK  CCAAEDQAAC  FTEEAPKLVS

ESAEIVKV
```

Figure 4.16. Amino acid sequence of Serum albumin-1 (complete) from Atlantic salmon *Salmo salar*. Length 608 AA, Mass 67,151 Da. UniProtKB/Swiss-Prot P21848 (ALBU1_SALSA).

Similar to mammal albumin*Salmonidae* albumins bind palmitic acid and, in contrast to them, they do not bind nickel (as albumins of other fishes) (Metcalf et al, 1998a). MM of Chinook salmon albumins is 65230 Da, in brook trout - 66960 Da (Metcalf et al, 1998b). The amino-acid sequences of their NH_2- terminal fragments are determined (Xu, Ding, 2005), the sequence of the first 15 amino acids of Chinook salmon and brook trout are similar (Metcalf et al, 1998a). The albumin of brook trout, in contrast to other *Salmonidae*, contains sialic acids (Metcalf et al, 1998b). This is atypical for albumins of reptiles and mammals. The detection of sialic acids in structure of albumin in the ancient group of *Salmonidae* fishes together with the jawless and the amphibians, allowed us to assume that initially the albumin was glycoprotein, but it lost this modification, probably, at separation of the reptiles from the amphibians (Metcalf et al, 1998b).

```
         10         20         30         40         50         60
MQWLSVCSLL VLLSVLSRSQ AQNQICTIFT EAKEDGFKSL ILVGLAQNLP DSTLGDLVPL
         70         80         90        100        110        120
IAEALAMGVK CCSDTPPEDC ERDVADLFQS AVCSSETLVE KNDLKMCCEK TAAERTHCFV
        130        140        150        160        170        180
DHKAKIPRDL SLKAELPAAD QCEDFKKDHK AFVGRFIFKF SKSNPMLPPH VVLAIAKGYG
        190        200        210        220        230        240
EVLTTCCGEA EAQTCFDTKK ATFQHAIAKR VAELKSLCIV HKKYGDRVVK AKKLVQYSQK
        250        260        270        280        290        300
MPQASFQEMA GMVDKIVATV APCCSGDMVT CMKERKTLVD EVCADESVLS RAAGLSACCK
        310        320        330        340        350        360
EDAVHRGSCV EAMKPDPKPD GLSEHYDVHA DIAAVCQTFT KTPDVAMGKL VYEISVRHPE
        370        380        390        400        410        420
SSQQVILRFA KEAEQALLQC CDMEDHAECV KTALAGSDID KKITDETDYY KKMCAAEAAV
        430        440        450        460        470        480
SDDNFEKSMM VYYTRIMPQA SFDQLHMVSE TVHDVLHACC KDEPGHFVLP CAEEKLTDAI
        490        500        510        520        530        540
DATCDDYDPS SINPHIAHCC NQSYSMRRHC ILAIQPDTEF TPPELDASSF HMGPELCTKD
        550        560        570        580        590        600
SKDLLLSGKK LLYGVVRHKT TITEDHLKTI STKYHTMKDK CCAAEDQAAC FTEEAPKLVS

ESAELVKV
```

Figure 4.17. Amino acid sequence of Serum albumin-2 (complete) from Atlantic salmon *Salmo salar*. Length 608 AA, Mass 67,059 Da. UniProtKB/Swiss-Prot Q03156 (ALBU2_SALSA).

```
         10         20         30         40         50         60
QNLPDSTLGD LVPLIAEAIA MGVKCCSDTP PEDCDRDVAD LFQSAVCSSE TLVEKNHLKM
         70         80         90        100        110        120
CCEKTAAERT HCFPDHKAKI PRDLSLKAEL PAADQCEDFK KDHKAFVGRF IFKFSKSNTM
        130        140        150        160        170        180
LQPHVILAIA KAYGEVLTSC CGEAEAQTCF DTKKATFQRA VGKRVTELRA LCIVHKKYGD
        190        200        210        220        230        240
RVVKAKKLIQ YSQKMPQASF QEMGGMVDKI VATVAPCCSG DMVTCMKERK ALVDEVCADK
        250        260        270        280
SVLSRAAGLS ACCKEDAVHR GSCVEAMKPD SKPDGLSEHY D
```

Figure 4.18. Amino acid sequence of Serum albumin-1 (fragment) from Rainbow trout *Salmo gairdneri*. Length 281 AA, Mass 30,750 Da. UniProtKB/TrEMBL: D5H440 (D5H440_ONCMY).

The Atlantic salmon gene **alb1** encodes the albumin (Serum of albumin 1) consisting of 608 amino-acid residues. The structure of protein is stabilized by S-S- bonds (UniProtKB/Swiss-Prot: P21848 (ALBU1_SALSA)); the gene alb2 encodes protein (Serum albumin 2) consisting of 608 amino-acid residues also (UniProtKB/Swiss-Prot:Q03156 (ALBU2_SALSA)).

The gene of serum albumin 1 encodes the protein of the Rainbow trout albumin, its fragment of 281 amino-acid residues is described (UniProtKB/TrEMBL: D5H440); the fragment of 167 amino-acid residues is also described (UniProtKB/TrEMBL: A5H0J9).

Family Siluridae

The representatives of another ancient unspecialized group of the fish - *Siluriformes* – have albumins also. The common bullhead *Ameiurus nebulosus* has serum albumin with MM approximately 68-70 kDa; its amino-acid composition, in contrast to albumins of other fishes, is characterized by the high content of histidine (6,7%) (Szebedinszky, Gilmour, 2002). There is no typical albumin in the blood serum of the fresh-water Channel catfish*Ictalurus punctatus*, but there is one protein, which has a higher affinity for zinc, than albumins of mammals and birds. Possibly, this protein is albumin (Bentley, 1991).

Family Anguuillidae

The typical albumins in another ancient group of fishes from the family *Anguillidae* are not discovered. In the blood plasma of two species of eels - *Anguilla dieffenbachia* and *Anguilla australis schmidtii* - the protein-lipoprotein is revealed. This protein has mobility of HSA in agarose; the autoradiography revealed the binding of the [^{14}C]-palmitic acid by this protein; the binding of $^{63}Ni^{2+}$ is not discovered (Metcalf et al, 1999a). The protein with the value of albumin MM among eels is not found. The palmitic acid is also connected by three additional proteins. One of them, most explicit on the electrophoregram, has MM approximately 30 kDa, that it does not coincide with MM of HSA too. The amino-acid sequences of the NH_2-terminal fragments of the eel palmitate- binding proteins proved to be homologous to apolipoprotein AI, which, as it is assumed, in the absence of albumin plays the role of the carrier of fatty acids (Metcalf et al, 1999b). The typical albumin is not discovered in *Anguilla japonica* (Nunomura, 1991). The C-reactive protein of *Anguilla japonica* in immunoelectrophoresis is revealed in the zone of albumin mobility. In native conditions its MM was 120 kDa and in presence of reducing agent - 24 kDa, in the immunoelectrophoresis this protein is revealed in the zone of the mobility of albumin.

Family Percidae

The two proteins from blood of *Perca fluviatilis* L. bind the dye Evans blue. It makes possible to consider these proteins as albumin-similar. MM of these proteins under the native conditions is approximately 120 kDa, in SDS-PAGE - about70 kDa (in the presence of reducing agent) (Andreeva, 2008).

Family Nototheniidae

There is no evidence for the presence of typical albumins in the blood of *Nototheniidae*. There is one protein in the blood of Antarctic fish*Dissostichus mawsoni*, which is described as the protein, binding [^{14}C] - palmitic acid, but in contrast to HAS, not binding $^{63}Ni^{2+}$ (Metcalf et al, 1999b). This protein is turned to be the lipoprotein, which is decomposed in 2D-electrophoresis into three small proteins with MM 11, 30 and 42 kDa, without any traces of protein with MM of HSA. The high level of homology of NH_2-terminal sequence fragment of the palmitate-binding protein to mammalian apolipoprotein AI (the basic apolipoprotein of the high density) is established. The other proteins are discovered also, which bind $^{63}Ni^{2+}$, but don't bind palmitic acid, and do not have homology with mammalian albumins (Metcalf et al, 1999b).

Family Mullidae, Uranoscopidae, Gobiidae

The Red mullet *Mullus barbatus*, Stargazer *Uranoscopus scaber*, Toad goby *Mesogobius batrachocephalus* P. and Round goby*Neogobius melanostomus* P. have blood plasma proteins, which bind Evans blue. They are identified as albumin-similar proteins (Andreeva et al., in of publ.).

Family Scorpaenidae

The five serum proteins with MM from 64 to 70 kDa (2D-SDS-PAGE), binding Evans blue, are revealed in the Black Scorpionfish *Scorpaena porcus* L. blood. MALDI-TOF-analysis of these proteins revealed the identity of their mass-spectra (MS), that indicates to their homology (Andreeva,2011).The search of the homologues for these proteins was carried out in the NCBI database among the proteins of all organisms. The total search with the use ofMS+MS/MS was performed with the BioTools v.3 (Bruker, Germany) program, however, reliable homologues were not discovered. The binding of albumin-specific dye by the proteins and the low MM values make it possible to consider these proteins as the albumin-similar ones (Andreeva,2011).

Family Cyprinidae

The complex albumin system is discovered in the fresh-water **Cyprinidae** and **Percidae** fishes. The basic component of this system is albumin-similar protein, which has MM approximately 67 kDa in the SDS- electrophoresis, it is glycoprotein (Andreeva, 1999, 2010a, 2010b). The Far Est redfin *Tribolodon brandti* Dybowskii, inhabiting the Sea of Japan and the Okhotsk Sea, is just one *Cyprinidae* species, who can successfully live both in fresh and saline waters. The redfin albumin is very similar to the bream albumin (Andreeva, 2010b). The albumin-similar protein is discovered in the blood of the Common carp *Cyprinus carpio*, its MM and isoelectric point differ from HSA (De Smet et al, 1998).

Family Clupeidae

The serum protein, like the albumin-similar protein of *Cyprinidae* is discovered in the blood of the Black and Caspian sea kilka *Clupeonella cultriventris* N. (Andreeva, 2008, 2010b).

Family Gadidae and Labridae

The blood plasma of shore rockling *Gaidropsarus mediterraneus* L. and labrid *Symphodus ocellatus* F. contains the proteins, which binds albuminspecific dye Evans blue. This makes it possible to consider these proteins as the albumin-similar proteins (Andreeva et al., in of publ.).

Thus, proteins that are diverse in their physical and chemistry properties are revealed in the blood of bony fishes. These proteins are similar (to one degree or another) or not similar to albumins of the highest vertebrates.

4.5. STRUCTURAL ORGANIZATION OF LOW-MOLECULAR PLASMA PROTEIN FRACTION OF FRESH-WATER *TELEOSTEI*

Serum low-molecular proteins and serum albumin in the blood of fresh water bony fishes form the dynamic fraction, which we conditionally named as the low-molecular fraction (LMF). The conditionality is in the fact that some components of this fraction have sufficiently high MM values (about 120-160 kDa) under the nondenaturing conditions of concentration PAGE gradient (3-40%). Two subfractions - albumins and low-molecular proteins are separated in this fraction clearly (Figure 4.19).

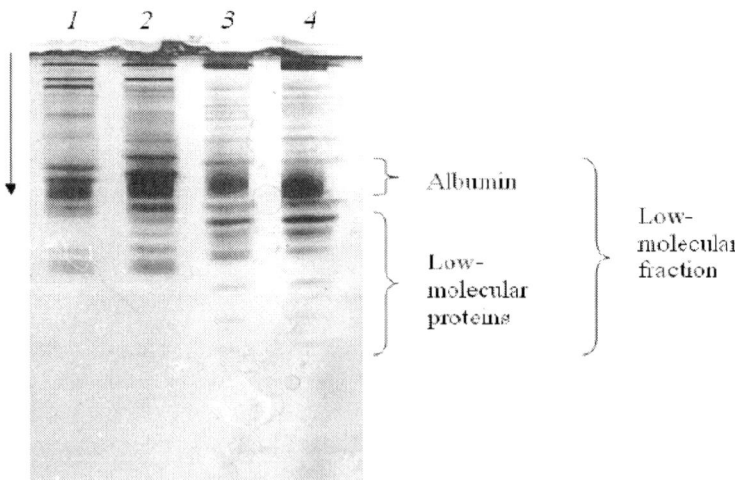

Figure 4.19. Disk-electrophoresis of bream (*1, 2*) and pike perch (*3, 4*) blood serum proteins. Vertical arrow show the direction of electrophoresis, curly brackets show zones of low-molecular fraction and its subfractions – "albumin" and "low-molecular proteins". (Andreeva, 1999).

Their combination in the composition of a single fraction is not occasional - both subfractions are reconstructed in the process of the plastic exchange and during the fish adaptations to the different water salinity (Chapter 6). The described types of organization of the low-molecular fraction are characteristic both for underyearling, young fishes and for the mature ones.

Subfraction "Low-Molecular Proteins"

The low-molecular subfraction of blood serumconsist of from 1-2 to 10 proteins. In summer in mature bream this subfraction consists of 5-6 components on the electrophoregram, which can be grouped into three groups - A, B and C, on 1-2 components in each (Figure 4.20).

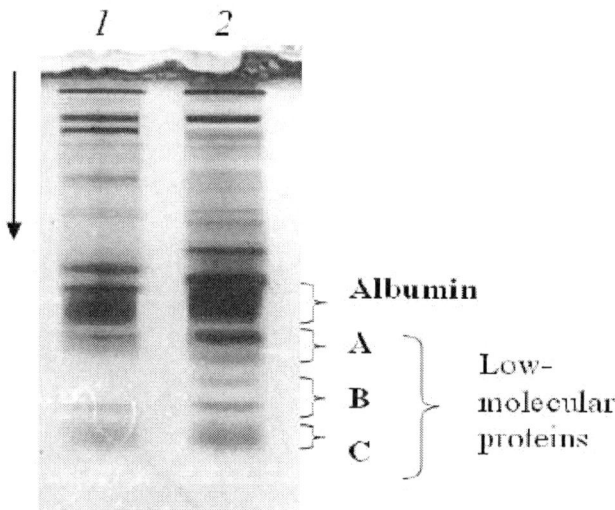

Figure 4.20. Disk-electrophoresis of bream blood serum proteins (*1, 2*). Vertical arrow show the direction of electrophoresis, curly brackets show zones of subfractions "albumin" and "low-molecular proteins"; A, B, C – components of subfraction "low-molecular proteins". (Andreeva, 1999).

The analysis of the character of their variability and mobility in the electrophoresis allow us to assume that their gene- determination occurs by three independent loci (Andreeva, 1999). Under native conditions these proteins had different MM values - from 80 to 96 kDa, in SDS-PAGE the heterogeneity disappeares and proteins has MM approximately 54 kDa. More likely, the proteins of this subfraction are monomers.

The quantity and the character of the dyeing of protein bands in probable heterozygotes on the locus B also testify in favor of the monomeric albumin structure, encoded by locus B. All proteins of this subfraction contained carbohydrates in their structure (Andreeva, 1999).

Subfraction "Albumins"

As for the albumin subfraction, it is located above low-molecular proteins subfraction on the electrophoregrams and have different MM values during different periods of the reproductive cycle: from 120-160 to 90 kDa (under natural conditions) (Andreeva, 1999) (Figure 4.21).

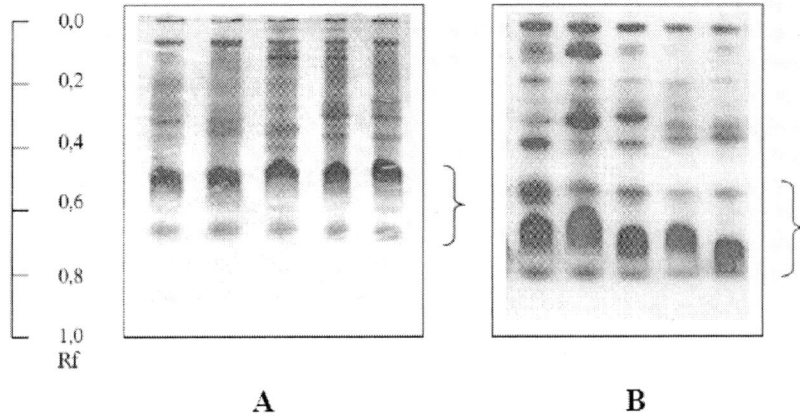

Figure 4.21. The types of organization of low-molecular fraction of serum proteins of fresh-water bony fishes – basic (A) and plastic (B) ones. Rf - electrophoretic mobility; curly bracket marks out zone of low-molecular fraction in PAGE (disk-electrophoresis). (Andreeva, 1999).

These proteins bind albumin-specific dye – Evans blue, bromphenol blue and bromcresol purple. They form the nonspecific complex with the latter in contrast to HSA (Andreeva, 1986). Both proteins have MM approximately 67 kDa in SDS-PAGE and contain carbohydrate in the molecule structure (Andreeva, 1999). Native proteins are lipoproteins (Andreeva, 1997).

Basic and Plastic Types of the Organization of Low-Molecular Fraction (LMF)

There are some typical methods of organization of serum protein low-molecular fraction in fresh-water bony fishes – basic and plastic ones (Figure 4.21).

The Basic Type of the Organization of Low-Molecular Fraction
The main time of the annual cycle subfraction "albumins" is represented by protein with R_f 0,55 and MM (the native conditions) about 120-130 kDa; and subfraction "low-molecular proteins" is maximally heterogeneous (from 2 to 10 components on the electrophoregram), maximum value R_f for the most mobile component is 0,63. This type of the organization of low-molecular fraction on the electrophoregram we named as "basic" (Figure 4.21).

The Plastic Type of the Organization of Low-Molecular Fraction
The second type of the organization of low-molecular fraction was observed in spring on the eve of the spawning in fishes with the gonads of the maturity stage IV-V (Figures 4.21, 4.22).

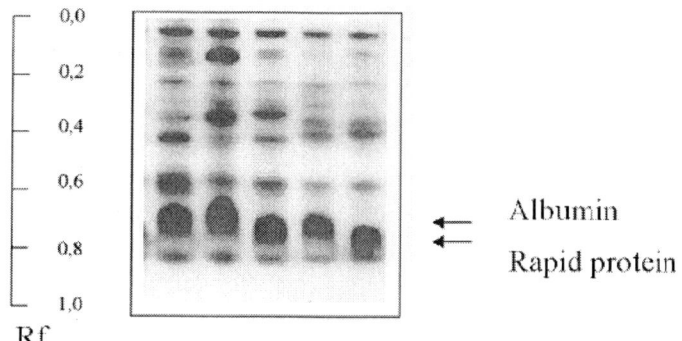

Figure 4.22. Plastic type of low-molecular fraction organization of bream serum proteins (disk-electrophoresis). Albumin and Rapid protein are components of low-molecular fraction of serum proteins, which is organizedby plastic type. Explanation is in the text. (Andreeva, 1999).

It is characteristics for this type of organization: subfraction "albumins" are represented by protein with $R_f = 0,7$ and MM about 90 kDa(the native conditions), while the subfraction "low-molecular proteins" is represented always by one protein with $R_f = 0,83$. We named this protein as "rapid" (Figures 4.21, 4.22).

Such type of the distribution of the proteins from low-molecular fraction on the electrophoregram we named "plastic", since it characterizes the high level of metabolic processes in the organism in the prespawning period, in which albumins take active part as the plastic material - the source of amino acids (Novikov, 2000).

The analysis of two types of organization of the low-molecular fraction has demonstrated that the rapid protein can be the product of the albumin degradation. However, MALDI-TOF-analysis shows the complete noncoincidence of the mass-spectra MS and fragmentation mass-spectra MS/MS of albumin and rapid protein. 35 and 34 fragments of the tryptic cleavage of the albumin and rapid protein respectively were obtained and there is not any pairs of fragments, coincident in MM value in these two proteins (Table 4.1).

Table 4.1. The values of Molecular weight (MM) of tryptic cleavage fragments of the proteins from bream blood plasma low-molecular fraction LMF

The name of protein	The name of subfraction of the blood plasma proteins	MM of protein fragments after tryptic cleavage (Da)
Albumin	"albumin"	806,45; 893.38; 909.49; 991.46; 1026.59; 1060.59; 1100.52; 1116.64; 1151.62; 1199.63; 1221.57; 1263.03; 1270.57; 1587.84; 1719.86; 1787.98; 1794.83; 1886.85; 1903.95; 1908.93; 1920.87; 1980.95; 1991.89; 1996.85; 2364.04; 2410.04; 2415.02; 2426.09; 2465.20; 2474.18; 2481.13; 2506.01; 2522.04; 2544.01; 2903.87
Rapid protein	LMP	866,46; 874,55; 946,61; 1050,59; 1069,68; 1085,71; 1101,68; 1124,66; 1182,66; 1206,66; 1235,73; 1238,67; 1257,68; 1305,69; 1411,79; 1433,76; 1456,75; 1611,89; 1711,86; 1814,87; 1829,90; 1852,86; 2042,99; 2059,02; 2061,02; 2075,01; 2080,99; 2397,24; 2412,26; 2444,19; 2459,20; 2475,20; 2743,32; 3285,62

The obtained results testify to the absence of the relationship between the analyzed proteins. Therefore the rapid protein cannot be the product of the albumin degradation. This result shows that the low-molecular fraction consists of two autonomous subfractions, that contain nonhomologous proteins.

4.6. THE OLYGOMERIC AND MONOMERIC PROTEINS IN THE COMPOSITION OF LOW-MOLECULAR BLOOD PROTEIN FRACTION OF FRESH-WATER BONY FISHES

The Oligomeric Proteins in the Fresh-Water Fishes

Theoligomeric albumins are found in the composition of blood of low-molecular fraction of the fresh-water bony fishes: the pike *Esox lucius* L, the zope *Abramis ballerus* L., the bream *Abramis brama* L., the ablet *Alburnus alburnus* L., the silver bream *Blicca bjoerkna* L., the ide *Leuciscus idus* L., the sabrefish *Pelecus cultratus* L., the roach *Rutilus rutilus* L., the crucian carp *Carassius carassius* L. and the goldfish *Carassius auratus* L., the common carp *Cyprinus carpio* L., the perch *Perca fluviatilis* L., the zander *Stizostedion lucioperca* L. and the Volga zander *St. volgense* G. et al.

The Basic Type of the Low-Molecular Fraction
10-13 proteins are discovered in the composition of the oligomeric albumin of the basic type in fish blood serum. So many proteins are revealed on the path of olygomeric albumin in PAGE with 8M urea (Figure 4.23). So many subunits are revealed both in the composition of albumin and in SDS-PAGE under reducing and nonreducing conditions (Figure 4.23). So the oligomeric albumin consist of 10-13 proteins, which are found by noncovalent bonds into the protein complex.

The Plastic Type of the Low-Molecular Fraction
Two proteins with MM approximately 13 and 27 kDa are discovered in the composition of the oligomeric albumin of plastic type in fish blood serum(Figure 4.24). These proteins are the products of the proteolysis of protein with MM 67 kDa (Andreeva, 2010b). Thus, the albumin and products

of its partial proteolytic degradation enter in the composition of the plastic type oligomeric albumin.

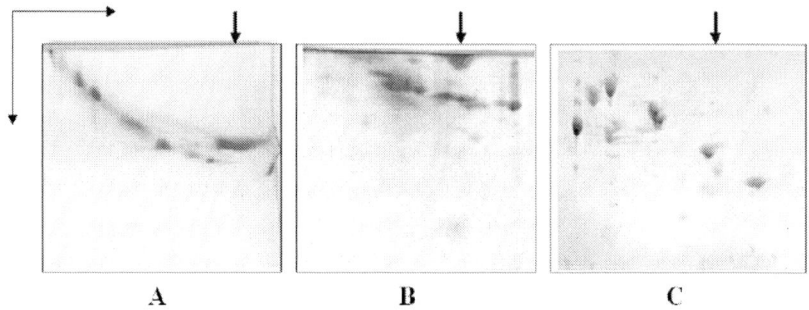

Figure 4.23. Eelectrophoregrams of serum proteins of the bream in PAGE (basic type): in gradient of PAGE concentration 4-30% electrophoresis (A), in PAGE with 8M urea (B) and in SDS-electrophoresis in reducing conditions (C). Vertical arrow show the direction of electrophoresis in gradient PAGE concentration (A), PAGE with urea (B) and SDS-PAGE (C); horizontal arrow - the direction of disk-electrophoresis. Little pointers indicates a track of albumin. (Andreeva, 2008).

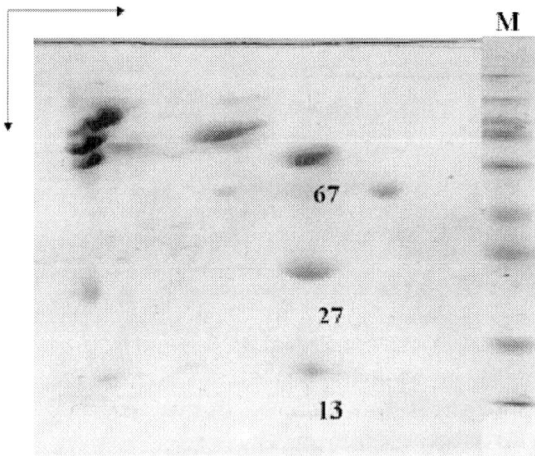

Figure 4.24. 2D-SDS-electrophoresis(reducing conditions) of serum proteins from roach (plastic type). Vertical arrow show the direction of SDS-electrophoresis, horizontal arrow - the direction of disk-electrophoresis. M – PageRuler[TM] Prestained Protein Ladder Plus Marker (Fermentas). 67, 27 and 13 – MM values in *kDa* of albumin and products its proteolysis. (Andreeva, 2010b).

4.7. OLYGOMERIC AND MONOMERIC PROTEINS IN THE COMPOSITION OF LMF BLOOD PROTEINS OF SALTISH WATER KILKA *CLUPEONELLA CULTRIVENTRIS* N. AND MIGRATORY SPECIES *TRIBOLODON BRANDTII* D

The olygomeric albumins are discovered in the subfraction "albumins" in the blood plasma of kilka *Clupeonella cultriventris* N. (the Rybinsk Reservoir) and *Tribolodon brandtii* D. (the Sea of Japan). These olygomeric albumins dissociated to 6 (kilka) and 13 (tribolodon) components under the conditions of 8M urea in PAGE (Andreeva, 2010b).

4.8. THE ORGANIZATION OF LOW-MOLECULAR FRACTION OF BLOOD PLASMA PROTEINS FROM MARINE BONY FISHES

The Differentiation of the Proteins from Low-Molecular Fraction of the Blood Plasma Proteins by the Charge and MM Values

The marine fishes – the shorthorn sculpin *Myoxocephalus scorpius* L., the sole *Liopsetta glacialis* P., the atlantic cod *Gadus morhua* L.; the round goby *Neogobius melanostomus* P., the goad goby *Mesogobius batrachocephalus* P., the horse mackerel *Trachurusmediterraneus* S., the pickerel *Spicara flexuosa* R., latterino sardaro *Atherina hepsetus* L., the golden grey mullet *Lisa aurata* R., the whiting *Merlangus merlangus euxinus* N., the shore rockling *Gaidropsarus mediterraneus* L., the corkwing *Symphodus tinca* L., the goatfish *Mullus barbatus*, the scorpionfish *Scorpenaporcus* L., the stargazer *Uranoscopus scaber* – have highly heterogeneous plasma low-molecular fractions. There are about 10 proteins with MM values from 20 to 90 kDa are revealed in their composition in the disk- electrophoresis (Figure 4.25).

The protein number from low-molecular fraction reaches 12-18 in 2D-electrophoresis in the gradient of PAGE concentration (5-40%); and this number is comparable in denaturing (PAGE with 8M urea) and not denaturing (PAGE without the urea) conditions with the exception of scorpionfish. The scorpionfishprotein number in the PAGE with 8M urea considerablyexceeded one in the PAGE without urea(Table 4.2; Figure 4.26) (Andreeva, 2011).

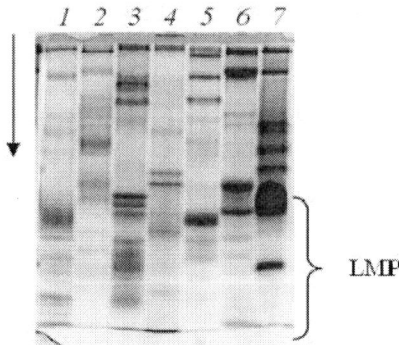

Figure 4.25. Disk- electrophoresis of the blood plasma proteins of goatfish *Mullus barbatus* L. (1), shore rockling*Gaidropsarus mediterraneus* L. (2), toad goby *Mesogobius batrachocephalus* P. (3) and goad goby *Neogobius melanostomus* P. (4), *Uranoscopus scaber* L. (5), scorpaena *Scorpaena porcus*(6) and roach *Rutilus rutilus*(7).LMP - low-molecular proteins. Vertical arrow shows the electrophoresis direction. (Andreeva, 2011).

Table 4.2. Maximal number of low-molecular proteins LMP of blood plasma of marine bony fishes in 1D и 2D-electrophoresisin native (disk-, PAGE concentration gradient 3-40%) and denaturative (PAGE with 8M urea, SDS-PAGEin reducing conditions) conditions

The marine *Teleostei* 1	1D Disk PAGE 2	2D Gradient PAGE 3	2D Urea PAGE 4	2D SDS- PAGE 5
Goatfish	6	10	11	12
Stargazer	9	15	17	24
Corkwing	7	16	17	24
Shore rockling	8	18	20	21
Scorpionfish	7	15	24	34
Toad goby	7	12	11	22
Goad goby	10	18	16	41

All investigated fishes have 1-3 proteins with MM value in the range of 60-75 kDa (most close to MM of HSA) in the composition of the low-molecular fraction: the stargazer has three such proteins, the shore rockling and the gobies - two proteins, the corkwing and the goatfish - one protein with MM value approximately 65 kDa, the scorpionfish - two proteins with MM approximately 64 and 69 kDa, the atlantic cod and the shorthorn sculpin - one protein with MM approximately 70 kDa, the sole - two proteins with MM approximately 67 kDa (Andreeva, 2008).

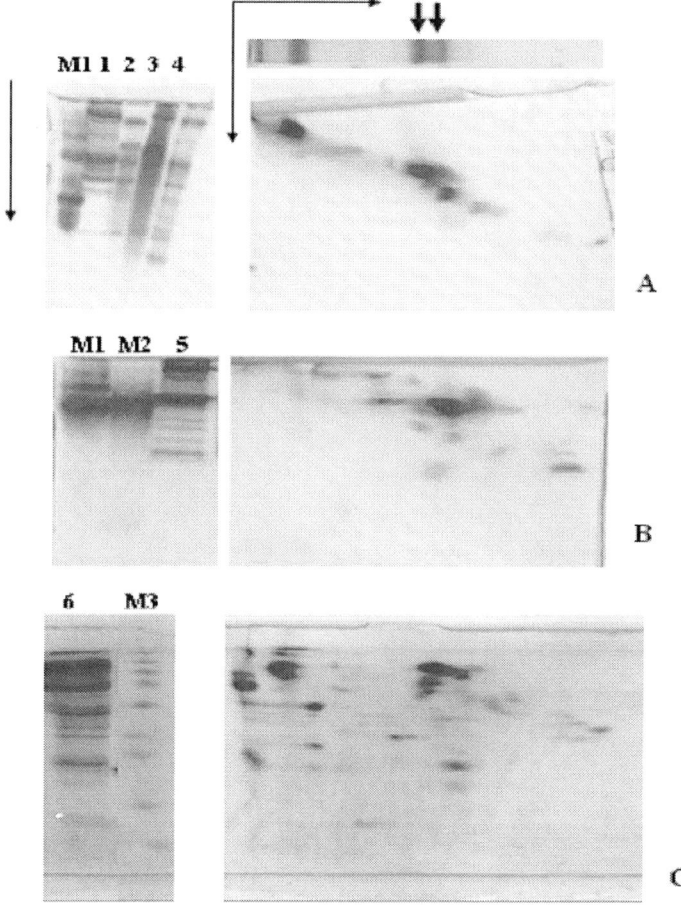

Figure 4.26. 2D-electrophoresis of plasma and tissue fluid proteins from scorpionfish: in the PAGE concentration gradient (**A**), in PAGE with 8M urea (**B**) and SDS-PAGE (**C**).1- scorpaena plasma; 2, 3, 4 - tissue fluids from peritoneal, white muscles and the brain.Marker proteins: M1 - HSA and OVA; M2 - myoglobin, M3 - PageRuler[TM] Prestained Protein Ladder Plus Marker (Fermentas).Horizontal arrow shows the disk-electrophoresis direction, vertical – gradient-electrophoresis, electrophoresis with urea and SDS-electrophoresis directions respectively. Two small vertical pointersindicates atracks of proteins with MM 60-70 kDa. (Andreeva, 2011).

The Binding of Low-Molecular Proteins by the Albumin-Specific Dyes

Spectrophotometric analysis has not revealed the formation of the specific complexes of bromocresol purple (BCP) with the proteins from extracellular fluids of marine fishes. Meanwhile this dye binds all roach serum proteins unspecifically, shifting λ_{max} from 590 to 593 nm (but not to 603 nm, which is characteristic for the specific binding) (Andreeva, 1985, 1987c). The binding of BCP by the marine fish proteins in PAGE has not been revealed during protein staining after electrophoresis, while all roach serum proteins are stained by BCP on the electrophoregram. Another albumin-specific dye Evans blue binds one plasma protein on the disk-electrophoregram by all marine fishes except the scorpionfish. Scorpaena plasma like fresh-water perch contains two proteins which bind Evans blue (Figure 4.27).

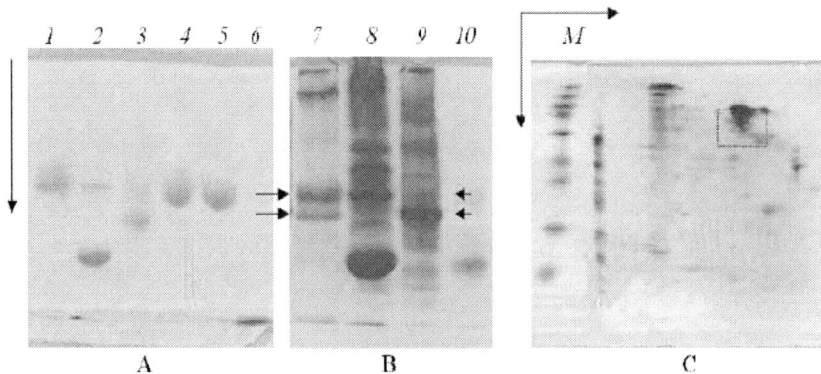

Figure 4.27. The binding of Evans blue by the bloodplasmaproteins from:

A - scorpaena (*1*), human (*2*), perch (*3*), HSA (*4*); controls: Evans blue (*5*), bromphenol blue (*6*) in the disk- electrophoresis;

B - the staining of proteins from scorpaena (*7*), human (*8*), perch (*9*) and HSA (*10*) by Coomassie R-250 in the disk-electrophoresis; small horizontal arrows show the areas of Evans blue binding; vertical arrow shows the disk-electrophoresis direction;

C – 2D-SDS-electrophoresis of scorpaena plasma proteins; the proteins, which bind Evans blue, are outlined by the frame. M – PageRuler™ Prestained Protein Ladder Plus Marker (Fermentas). Vertical arrow shows SDS-electrophoresis direction, horizontal –disk-electrophoresis direction. (Andreeva, 2011).

1-3 components, which bind Evans blue in the native 2D-PAGE on the paths of precisely these proteins are discovered. And namely these proteins have MM value in the range of 60-76 kDa in SDS-PAGE, which is the most close MM of HSA.

4.9. STRUCTURAL ORGANIZATION OF LOW-MOLECULAR BLOOD PROTEINS OF MARINE *TELEOSTEI*

The Search for Protein- Oligomers in the Blood of Sea Species

The protein- oligomers in the low-molecular fraction of the blood are not discovered in the overwhelming majority of marine fish species. The protein component with the illegible contours on the disk-electrophoregram is discovered in the serum of the shorthorn sculpin (Andreeva, 2008). However, this component was decomposed to several proteins in the concentration PAGE gradient (5-40%). Probably, the proteins of the shorthorn sculpin have a certain tendency toward the complexing, but participating in this weak intermolecular forces are insufficient for maintaining the stable protein complex.

The low-molecular proteins of the stargazer, the goatfish, the corkwing, the gobies and the shore rockling are predominantly monomeric, their quantities (proteins) on the 2D-electrophoregrams in the concentration PAGE gradient and in the PAGE with urea are compared. In this case the proteins of the stargazer, the gobies and the corkwing consist of several polypeptide chains bound by S-S- bridges, and proteins of the shore rockling and the goatfish - of one chain (Table 4.2). And only in the case of scorpaena the number of proteins in PAGE with urea considerably exceeds the same in the nondenaturing electrophoresis. This circumstance suggests the presence not only of monomer, but also of oligomeric proteins in the low-molecular blood fraction of scorpaena. Among the olygomeric proteins there are ones consist of several polypeptide chains bound by S-S- bridges (Figure 4.28).

Mass-Spectra MS of Scorpaena Albumins

Scorpaena proteins, which bind Evans blue in the disk-electrophoresis, in the 2D-SDS-electrophoresis are represented by the large number of protein bands, among which there are five macrocomponents, which bind the dye; three of them have MM value 64 (protein-1), 69 (protein-2) and 70 (protein-3) kDa.

These proteins are albumin-similar, because they bind albumin-specific dye and have MM most similar to HSA.Being stained by Evans blue in disk-electrophoresis these proteins retained the dye solidly in SDS-PAGE, which made it possible to see complexes protein-dye without the additional PAGE staining. MALDI-TOF-analysis of these proteins revealed practically the complete coincidence of their mass-spectra (Figure 4.29).

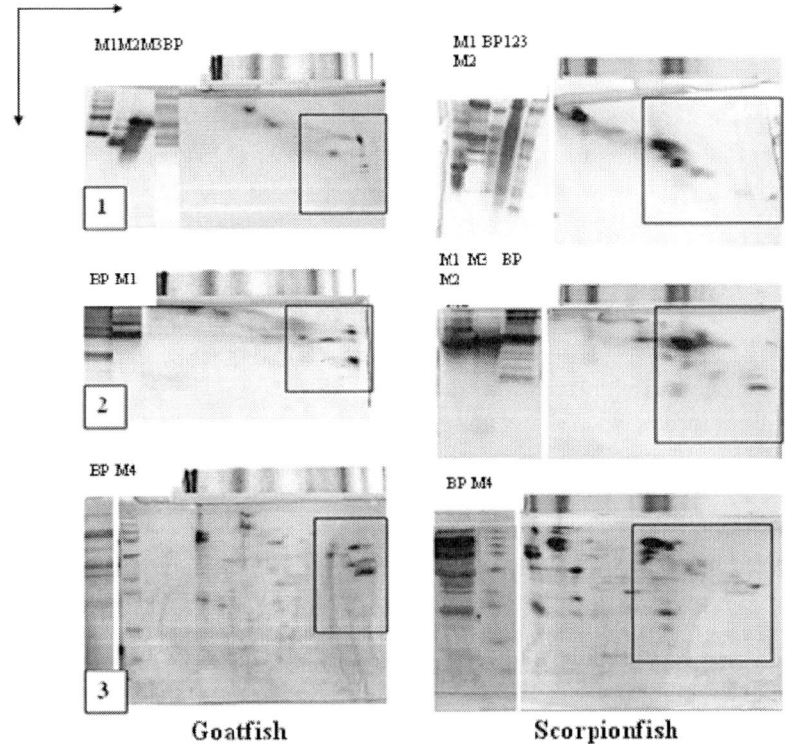

Figure 4.28. 2D-electrophoresis of plasma proteins of goatfish*Mullus barbatus*and scorpaena *Scorpaena porcus*in PAGE concentration gradient 4-30% (row 1), PAGE with 8M urea (row 2) and SDS-PAGE (row 3) electrophoresis. BP – blood plasma; M1 – HSA, M2 – OVA, M3 – myoglobin; 1-3 – scorpaena tissue fluids. M4 - PageRuler[TM] Prestained Protein Ladder Plus Marker (Fermentas). Vertical arrow shows direction of PAGE gradient, PAGE with urea and SDS-PAGE electrophoresis, horizontal –disk-electrophoresis direction. Low-molecular fraction of plasma proteins is framed.

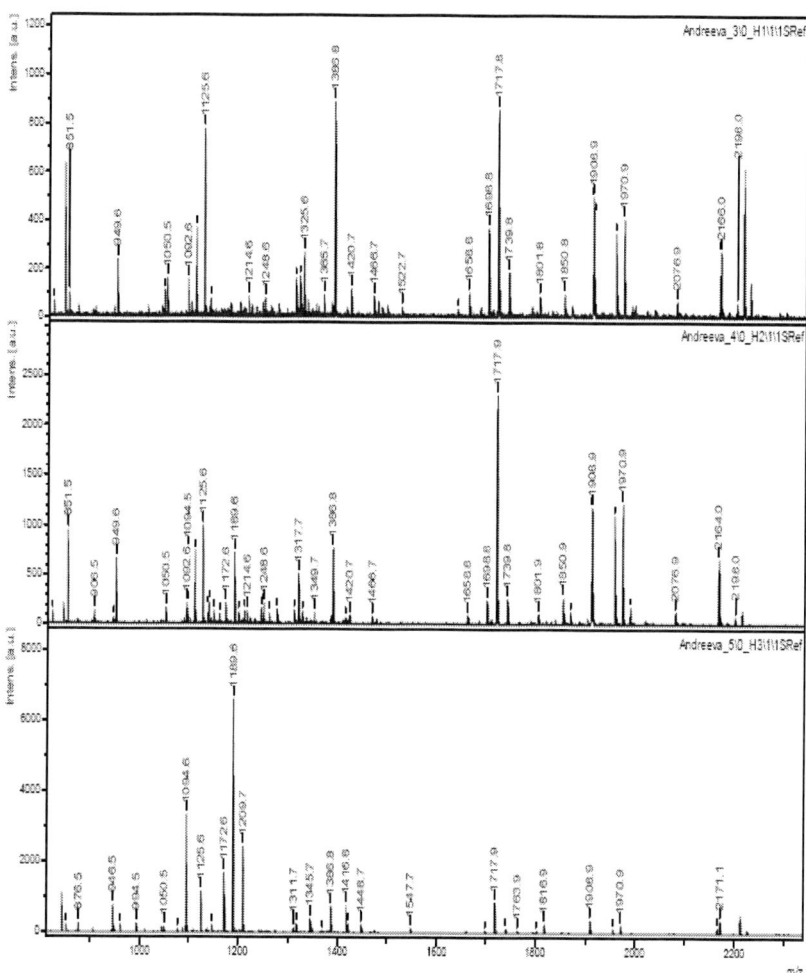

Figure 4.29. Mass-spectra of scorpaena serum albumins. The abscissa axis is the values of tryptic cleavage fragments MM, the ordinate axis is the signal intensity.

On the Table 1 the MM values of tryptic cleavage products of two albumin-similar proteins are presented. These proteins are differed only by three fragments (Table 4.3).

Table 4.3. MM of tryptic cleavage productsof scorpaena albumins

Component of fraction of albumin	MM of products after albumin tryptic cleavage, Da
1	723.38; 733.32; 778.40; 851.49; 901.43; 927.41; 949.57; 1011.51; 1050.53; 1083.52; 1125.62; 1197.59; 1214.65; 1243.6; 1248.57; 1256.64; 1309.63; 1310.60; 1317.73; 1386.78; 1420.69; 1466.74; 1509.80; 1636.81; 1658.83; 1660.69; 1680.86; 1682.79; 1698.78; 1704.88; 1717.86; 1739.85; 1785.89; 1808.95; 1866.83; 1908.86; 1954.88; 2165.02; 2540.06; 2629.26; 2805.20; 2933.31;3054.37;3121.7*;3399.54
2	723.38; 733.32; 778.40; 851.49; 901.43; 927.41; 949.57; 1011.51; 1050.53; 1083.52; 1125.62; 1197.59; 1214.65; 1243.6; 1248.57; 1256.64; 1309.63; 1310.60; 1317.73; 1386.78; 1420.69; 1466.74; 1509.80; 1636.81; 1658.83; 1660.69; 1680.86; 1682.79; 1698.78; 1704.88; 1717.86; 1739.85; 1785.89; 1808.95; 1866.83; 1908.86; 1954.88; 2165.02; 2540.06; 2629.26; 2805.20; 2933.31;3005.54*;3054.37;3070.37*;3399.54

*albumin tryptic cleavage products, which MM doesn't match. (Andreeva, 2011).

These data show the plurality of scorpaena albumin. It can be the consequence of two differentmechanisms. I - it is possible to assume that these albumin-similar proteins are the products of different genes, which are united by the same origin. The detection of the identical amino-acid fragments in these proteins can be explained by duplication of initial (ancestral) gene. The presence of the amino-acid fragments in one protein, while they are absent in other protein, can arise from subsequent intra-genetic reconstructions - deletions or insertions. II - in this case the plurality of albumins is the consequence of posttranslational modifications or partial proteolysis of one protein. The albumin-similar proteins are obtained from the fishes with the gonads of IV-V stages of maturity. Exactly at these stages active catabolism of fish albumins occurs, and therefore, the variety of albumins in MM value can be the consequence of their partial proteolysis. In this case, the scorpaena oligomeric protein consists of albumin and products of its partial proteolysis.

The search for the homologues of these scorpaena proteins in the NCBI database gave no results, the reliable candidate proteins were not detected. However, these data allow us to assume that the marine fish blood plasma contains the albumin-similar proteins with MM, close to mammalian serum albumin. The surface structure of these proteins differes from mammalian albumins. Similar to HSA these proteins bind some albumin-specific dyes (Evans blue) and, in contrast to HSA, they don't bind the other dyes (BCP). Only one species of marine fishes have olygomeric albumin among serum proteins, and gomologous proteins enter the composition of this complex. There are no such olygomeric proteins in the blood of other marine *Teleostei*.

Thus, the analysis of the structural organization of the blood plasma proteins in bony fishes showed the presence of oligomeric protein in the low-molecular fraction only in fresh water and bracken water fishes. These oligomers consist of 6 and more proteins, which are bound each other by noncovalent bonds. Among marine *Teleostei* there are species, which have oligomeric protein in the low-molecular fraction (for example, Scorpionfish), and species which have no such proteins. The second include the majority of the fishes.

What can be testified by the presence or the absence of oligomeric proteins in the plasma low-molecular fraction- will be considered in following chapters (chapters 6, 8).

Chapter 5

DISTRIBUTION OF BLOOD PLASMA PROTEINS BETWEEN INTRAVASCULAR AND INTERSTITIAL FLUIDS

5.1. PROTEIN CONTENT IN THE INTERSTITIAL FLUIDS OF FISH

The most part of extracellular fluid in the body of the vertebrate is the tissue fluid. This fluid is frequently reviewed as the analog of the interstitial fluid. All interstitial fluids of fishes contain high protein concentrations (Table 5.1).

The protein concentration in interstitial fluids varies depending on age, physiological state of fishes and environmental factors. Thus, the protein concentration in blood plasma and interstitial fluidof white musclesof mature carp is 6.05 and 4.45 %, and in underyearlins – 4.8 and 1.9 %, respectively (Andreeva et al., 2007). In 2+ breams kept in aquariums the blood plasma protein concentration decreased almost 5 times after 8 months of starvation as compared with the average protein concentration in breams from foraging ponds, and in interstitial fluids from peritoneum and white muscles protein remained in trace amount, excluding the brain interstitial fluid, protein concentration in which was 0.7 %, that is also lower than the same of fish from foraging ponds (1.65 %) (Fedorov, Andreeva, 2009).

The high protein content in fish interstitial fluids is caused by high permeability of fish capillary walls to the plasma proteins (Hargens et al., 1974; Kiernan, Contestabile, 1980; Tsvetnenko, 1986; Chalov, Lukjanenko, 1989; Olson et al., 2003; Phillips, 2003; Andreeva et al., 2007, 2008;

Andreeva, 2010a; Andreeva, Fedorov, 2010). Functional suitability of such high permeability is, presumably, in the need of fish to endure the intravascular fluid volume fluctuations, in particular hypovolemia, and to facilitate the restoration of the intravascular fluid volume after hypovolemia (Duff, Olson, 1989).

Table 5.1. Protein concentration (%) in fish extracellular fluids

Fishes (individual)	Plasma	Serum	Peritoneal fluid	Brain fluid	White muscles fluid
Buckler skate		2,4	0,55		
Common stingray		0,47	1,15		
Sterlet (3+)	3,7	3,5	0,6		
Bream	6.1	5.7			4.5
Bream		4.9		2.65	
Goldfish		3,0	4,05	2,5	
Tench		3,2		1,85	
Roach		3,33		3,6	1,3
Roach		3,7	2,9	5,25	
Carp		2,75	2,1	1,65	0,95
Carp	6.05				4.45
Pike		3,1	1,95	1,5	
Pickarel		4,0	2,5		
Goad goby		1,85	0,65	1,0	
Stargazer		1,8		3,0	
Scorpionfish		2,75	1,5	2,0	

(From Fedorov, Andreeva, 2009, 2010; Andreeva et al., 2010).

5.2. FRACTIONAL COMPOSITION OF THE INTERSTITIAL FLUID PROTEINS IN FISHES

The protein fractional compositionof all tissue fluids of the fishes (peritoneal, brain fluid, white muscle fluid etc.) is similar to that of blood plasma. This is confirmed by the matching of protein subunit repertoire in all extracellular fluids (Figure 5.1). This is explained by the fact that all tissue fluids are filtrates of blood plasma.

But higher relative content of low-molecular proteins is recorded in the interstitial fluids of bony fishes as compared with blood plasma. The relative content of low-molecular proteins in tissue fluid from white muscles of carp underyearlins is 61.2 %, and in blood plasma – only 3.9% (Andreeva et al., 2007).The relative content of low-molecular proteins of the marine species –

scorpaena - in peritoneal fluid is 39.9%, in the brain interstitial fluid is 22.3% and in blood serum is 28%.

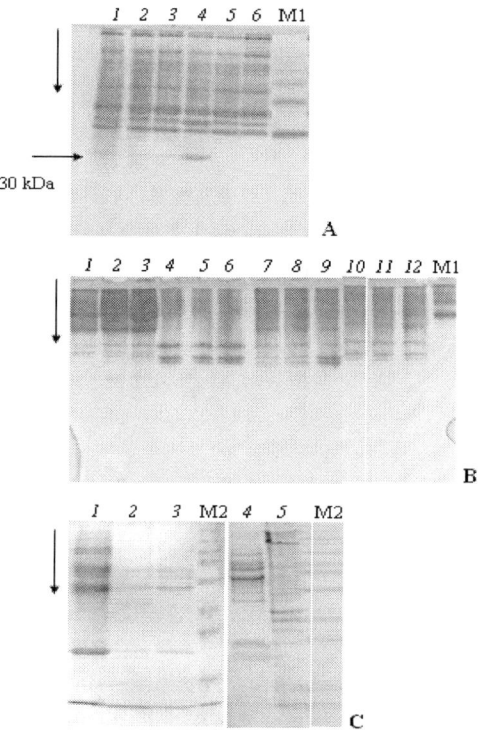

Figure 5.1. Electrophoresis of bream plasma and tissue proteinsin PAGE concentration gradient (A), PAGE with 8M urea (B) and SDS-PAGE (reducing conditions) (C):A – intestinal fluid (*1*), hepatic fluid (*2*), peritoneal fluid (*3*), white musclefluid (*4*), brain fluid (*5*), serum (*6*), M – HSA; B - serum (*1-3*), white musclefluid (*4-6*), brain fluid (*7-9*), peritoneal fluid (*10-12*),M - HSA; C- serum (*1*), whitemusclefluid (*2,5*),peritoneal fluid(*3*), blood plasma (*4*),M – PageRuler™ Prestained Protein Ladder Plus Marker (Fermentas).

Vertical arrow show the direction of electrophoresis, horizontal pointer indicate a location of protein with MM about 30 kDa. (Andreeva, 2010b).

A special low-molecular fraction with MM 20-40 kDa is found in interstitial fluids of bony fishes; it is presented in trace amounts in the blood serum (Figure 5.1) (Andreeva et al., 2007, 2008).Among its proteins 1-2 macrocomponents are usually registered: it is a protein with MM about 30 kDa under native conditions and 25 kDa under denaturing conditions in bream and roach (Figure 5.1A); two proteins with MM 45 and 30 kDa under native conditions in scorpaena; two by two macrocomponents with MM about 50 and 33 kDa under native conditions in polyploidic species – goldfish and carp (Andreeva et al., 2008). These components are the most abundant in interstitial fluid of white muscles.

5.3. DIFFERENTIAL PERMEABILITY OF DISTINCT SECTIONS OF CAPILLARY NETWORK TO THE BLOOD PLASMA PROTEINS

The protein concentrations can vary significantly in different interstitial fluids of fish. This is connected with differential permeability of different capillary network sections to the blood plasma proteins. Different capillary network sections are the most diametrically differentiated in higher vertebrates: from the complete impermeability of brain capillaries to the absolute permeability of capillaries in liver; capillaries of muscles and mesentery taking intermediate position (Chapter 1). The proteins reflection coefficient or the inverse to it filtration coefficient are used for estimating the vessel walls' permeability to the blood plasma proteins (Table 5.2).

Table 5.2. Filtration coefficients (r) of plasma proteins in different capillary sections of some fish

Fishes (individual)	Filtration coefficients (r)		
	Peritoneum	Brain	White muscles
Goldfish	0.69	0.58	
Goldfish	1.35	0.83	
Roach		1.08	0.39
Roach	0.78	1.42	
Carp	0.76	0.6	0.35
Pike	0.63	0.48	
Pike	0.85	0.63	
Goad goby	0.35	0.54	
Scorpionfish	0.55	0.73	
Scorpionfish	0.77	0.77	

The latter is calculated as the ratio of the protein concentration in the respective interstitial fluid to that in the blood plasma(Andreeva et al., 2007, 2008).

In spite of the fact that there are many cases of more high protein content in interstitial fluids as compared to the blood plasma, in most cases extravascular fluids of fish contain less protein than blood plasma. The plasma proteins concentration gradient through the capillary wall is maintained by the barrier mechanisms. The most well-studied of them is hematoencephalic barrier (HEB), which prevents penetration of proteins, blood cells and toxic compounds into the brain interstitial fluid. All vertebrates have HEB (Bundgaard, Cserr, 1981). In the most species it is formed by the endothelial cells of vascular wall, connected with each other by tight junctions. Only in elasmobranch fishes *Elasmobranchii* (sharks and skates) and in sturgeon fishes *Acipenseridae* HEB is formed by perivascular astrocytes. In mammals HEB is absent in six anatomic zones of brain, including capillaries of the pineal organ. The analysis of vessel permeability in the pineal body in rainbow trout *Salmo gairdneri* allows to assume the differentiation of blood vessels in this organ by the permeability to the compounds with different molecular weight (Omura et al., 1985). The features of similarity with mammals are revealed in HEB functioning in *Danio rerio* (Jeong et al., 2008). The HEB for hydrophilic molecules with MM above 900 Da is described in more primitive myxines, and the endothelial type HEB for macromolecules is described in lampreys (Bundgaard, 1982). The total protein content in the brain interstitial fluid of bream, roach, goldfish, tench, pike, carp, scorpionfish, gobies and other teleost fishes was comparable with that in blood plasma, however, filtration coefficients for individual plasma proteins differed. The reason of that is selective permeability of capillary walls for different plasma proteins.

5.4. SELECTIVE PERMEABILITY OF CAPILLARY WALLS FOR DIFFERENT BLOOD PLASMA PROTEINS OF FISHES

All fish interstitial fluids are blood plasma filtrates, however, the relative content of particular fractions and proteins in them is not the same (Figure 5.2).

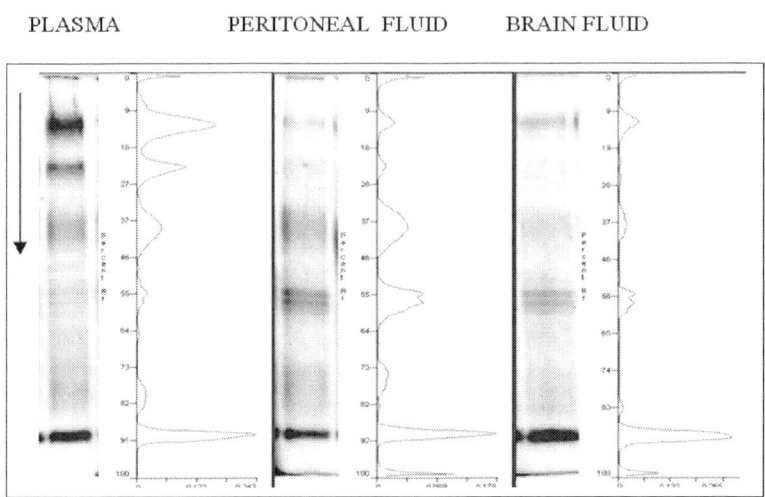

Figure 5.2. Electrophoresis of the proteins from blood plasma, peritoneal fluid and brain interstitial fluid of sterlet in PAGE concentration gradient (4-30%) (Fedorov, Andreeva, 2010).

The interstitial fluid of white muscles differs from the blood plasma most of all, its protein composition is the most dynamic and can differ significantly from the plasma protein fraction composition due to the active mechanisms, maintaining different blood proteins' gradient through the capillary wall. Nevertheless, the qualitative composition of subunits in blood plasma and white muscles matches completely (Andreeva, 2010b).

In sterlet the individual concentration gradient through the capillary walls is maintained for different proteins (Figure 5.3). Coefficients of filtration through the peritoneum capillaries varied from 0 to 1.7 for different sterlet plasma proteins, through the brain capillaries – from 0 to 0.8; average values were 0.73 and 0.39, respectively.

The analysis of filtration coefficients values for the proteins and their molecular weight values did not reveal any correlation between them. The sterlet capillaries forming interstitial fluid possessed comparable permeability for plasma proteins with MM from 65 to 500 kDa, and not displaying the molecular sieve effect (Tsvetnenko, 1986; Fedorov, Andreeva, 2010); however, they displayed selective permeability for different blood proteins (Figure 5.3).

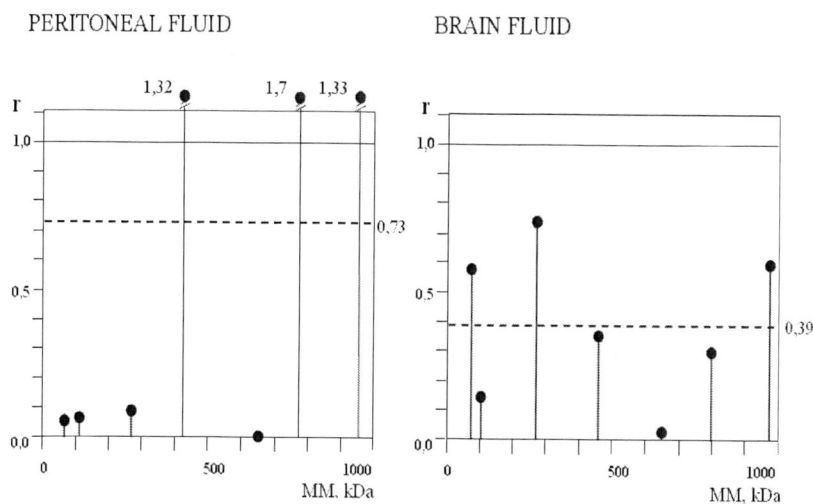

Figure 5.3. Coefficients of filtration (r) of sterlet blood plasma proteins to peritoneal and brain fluids (individual fishes) (Fedorov, Andreeva, 2010).

The same patterns were revealed in the analysis of blood protein filtration in other groups of fishes. For mature carp the filtration coefficients r_i for the wall of muscle-type capillary and average filtration coefficient for all proteins are calculated for different plasma proteins (Figure 5.4). Filtration coefficients r_i varied from 0.25 to 2.88 for different plasma proteins, average coefficient was 0.803. Eight proteins (1-8) were present only in the muscle interstitial fluid and were absent in blood plasma. The proteins *11* and *16* with different MM values (63 and 119 kDa) had almost equilibrium distribution between blood plasma and interstitial fluid of white muscles (r_i =1.05); for the most mobile component *9* with MM about 25 kDa r_i =2.88, and r_i =0.61 for the starting alpha-globulin *27*, which is comparable with r_i value for the low-molecular protein *13* with MM about 73 kDa (Figure 5.4). Therefore, the muscle-type capillary walls of the carp, like capillaries of sterlet, does not display molecular sieve effect, however, they possess selective permeability to different blood proteins.

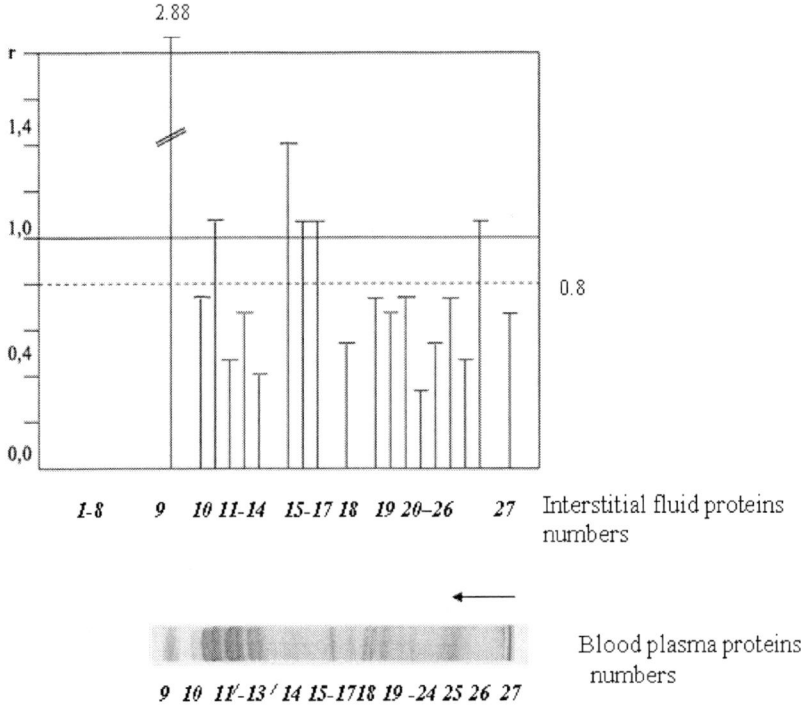

Figure 5.4. Coefficients of filtration (r) of blod plasma proteins of mature carp by capillaries of white muscles. 1-27 – numbers of plasma proteins in interstitial fluid and blood plasma. Horizontal arrow show the direction of PAGE concentration gradient (3-40%) electrophoresis. 1-8 -low-molecular proteins from interstitial fluid (Andreeva, 2008).

5.5. THE EFFECT OF DIFFERENT FACTORS TO TRANSCAPILLARY EXCHANGE OF BLOOD PLASMA PROTEINS

Permeability of capillaries is subjected to the effect of different factors, so, the filtration coefficient of proteins varies greatly in a single individual. The contentof such variability is in the adaptation of all metabolic processes of the organism to the changing environmental conditions.

Transcapillary Exchange of Blood Plasma Proteins in Sterlet under Heightened Salinity 20%

Under salinity the average coefficient of blood plasma proteins filtration into peritoneal fluidwas 0.29. For the control group of fish, which was kept in the aquarium with fresh water, the average filtration coefficient varied from 0.19 to 0.43 in different individuals. Thereby, the parameters of filtration of blood plasma proteins into interstitial fluid under high salinity remain in normal ranges.

In the control group the filtration coefficients for different proteins varied in a wide range: from 0.19 to 0.35 for albumins; from 0.17 to 0.38 for alpha-1-globulin (MM about 100 kDa); from 0.08 to 0.7 for oligomeric gamma-2-globulin (MM about 830 kDa); for transferrin coefficient values were close to 0. Under high salinity the values of the coefficient for individual proteins exceeded their normal fluctuations. Thus, the filtration coefficient for transferrin increased from 0 (control) to 0.28 under salinity, and for oligomeric gamma-2-globulin (MM about 830 kDa) it increased to 1.0; for albumin – it increased to the upper limit of norma (0.35), and for alpha-1-globulin – remained within the limits of norma (Figure 5.5).

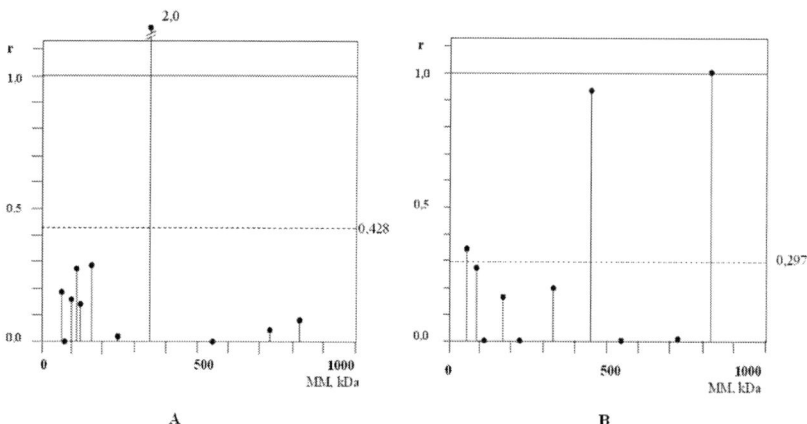

Figure 5.5. Coefficients of filtration (r) of sterlet blood plasma proteinsto peritoneal fluids in fresh (A) and saline (20 ‰) (B) water (individual fishes).

Thereby, total protein concentration in peritoneal fluid did not increase after placement of the nonmigratory fresh water sterlet from fresh into saline (20‰) water, however, plasma protein redistributionoccurred and the relative content of some proteins was changed:the relative content of transferrin and gamma-2-globulin increased, while the relative content of osmotic active protein albumin remained within the limits of norm.

Effect of Water Salinity on the Relative Content of Low-Molecular Proteins in Plasma and Interstitial Fluids of Teleost Fish

Scorpionfish Keeping in Desalinated Water

The relative content of low-molecular proteins in blood plasma of all individuals (20 fishes) was usually lower than in interstitial fluids. Keeping of scorpaena in salt and desalinated water did not change this trend in whole (Figure 5.6). The relative low-molecular fraction content values in experimental groups (dilution of salt water with fresh 3:1; fresh water) overlapped respective values in the control (salt water).

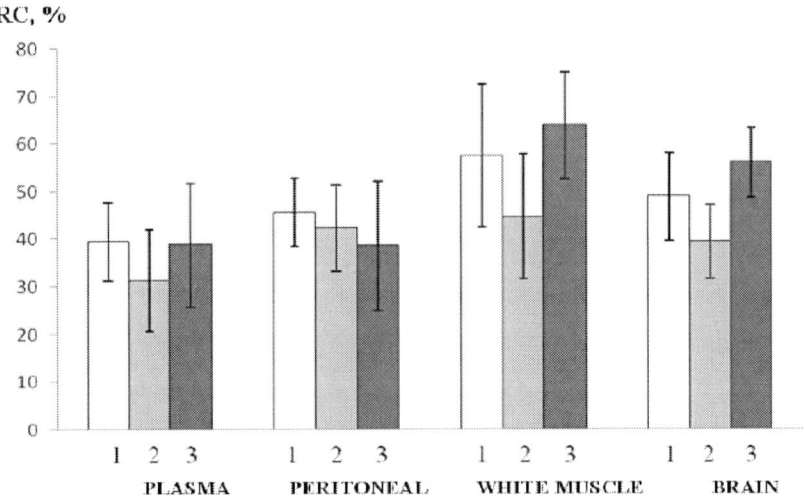

Figure 5.6. The relative content (RC, %) of low-molecular proteins in blood plasma and interstitial fluids (peritoneal, white muscle, brain) of scorpaena for fishes kept in salt water(1), diluted salt water 3:1 (2) and fresh water (3). (Andreeva et al., in of publ.).

Keeping Fresh Water Teleost Fishes in Salt Water

Meanwhile, for fresh water bream and roach redistribution of low-molecular fraction proteins, namely, the proteins from "low-molecular proteins" and "albumins" subfractions, occurred in all extracellular fluids during the adaptation to the salinity 8, 10 and 11.5‰ (Chapter 4) (Andreeva, 2010b).

Under increase of water salinity the relative content of "low-molecular proteins" subfraction in blood serum of fish remained almost the same, and in muscle interstitial fluid it decreased almost 9 times (Figure 5.7).

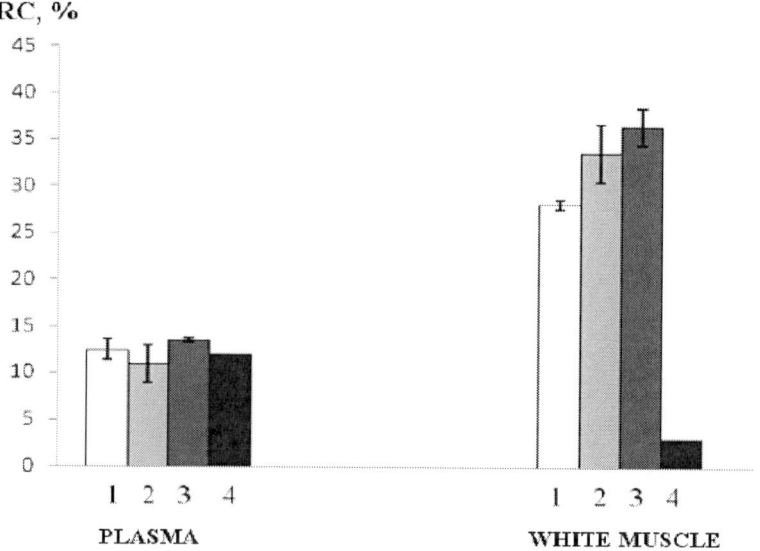

Figure 5.7. The relative content (RC, %) of proteins from the "low-molecular proteins" subfraction in blood serum and interstitial fluid of white muscles of breams kept in fresh (1) and saline water: 8‰ (2), 10‰ (3) and 11.5‰ (4). (Andreeva, 2010b).

The significant decrease of the relative content of oligomeric protein from "albumin" subfraction occurred during that (Figure 5.8). The decrease of the relative content of the abovementioned proteins in fish serum and tissue fluid in saline water concurred simultaneously with the increase of the proteins with MM 20-40 kDa in interstitial fluid (Figure 5.9). Such protein was absent in the blood serum.

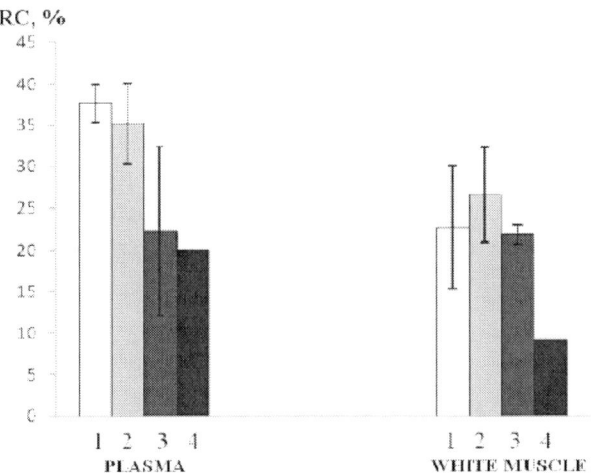

Figure 5.8. The relative content (RC, %) of "albumin" subfraction in serum and interstitial fluid of white muscles of bream kept in fresh (1) and saline water: 8‰ (2), 10‰ (3) and 11.5‰ (4). (Andreeva, 2010b).

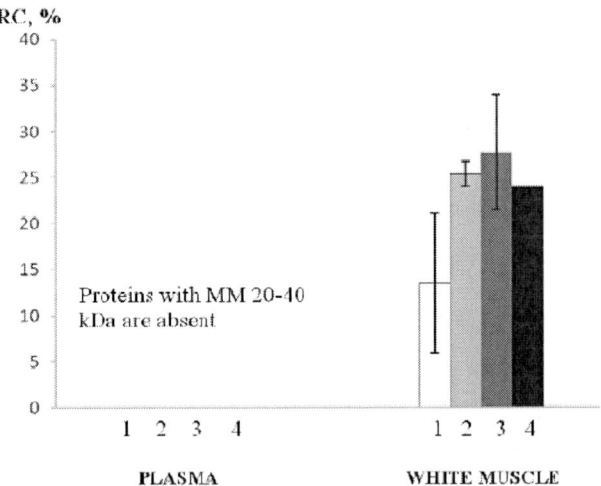

Figure 5.9. The relative content (RC, %) of proteins with MM 20-40 kDa in blood serum and tissue fluid of white muscles of breams kept in fresh (1) and saline water: 8‰ (2), 10 ‰ (3) and 11.5 ‰ (4). (Andreeva, 2010b).

Thereby, the relative content of low-molecular proteins in interstitial fluids of scorpaena and bream remained higher than that in blood plasma (serum) under changing water salinity. In the bream the redistribution of low-molecular proteins in the serum and interstitial fluid under saline water conditions was found. The mechanism of this redistribution will be considered in the next chapter.

The Effect of Food Supply on the Distribution of Interstitial Fluid in the Organism and on Permeability of Capillaries to the Proteins (by the Example of Sterlet and Bream)

The increase in the peritoneal capillaries permeability to plasma proteins and the increase of peritoneal fluid volume were recorded for sterlet (3+), which was kept in the pool with the excess of food for 7 weeks. During first days of fish keeping in the pool the average filtration coefficient of plasma proteins varied from 0.19 to 0.43 for different starlet individualy, after 7 weeks the "r" values increased to 0.74.

The increase in interstitial fluid volume relative to control was also recorded for breams (2+), which were kept in aquarium without food for 8 months, however, the protein was found in trace quantities in the peritoneal and white muscle fluids, and there was no correlation observed between the interstitial fluid volume increase and the protein concentration in it.

Thereby, when food is available the increase in the protein concentration in interstitial fluid is accompanied by the fluid volume increase (sterlet), and in the absence of food such a relationship is not observed (bream). Tissue watering during the decrease of protein concentration in interstitial fluids in the fishes, that were starving for a long time, can be explained by the emergence of non-protein low-molecular osmotic active molecules in interstitial fluids, such as, for example, amino acids. The albumin is used by the organism as a plastic raw material under starvation, exactly from which the amino acids are formed by proteolysis, then glucose molecules are formed from them during gluconeogenesis. Being osmotic active compounds, the molecules of amino acids and glucose forward to pump of extracellular fluid of the organism throughwall of capillary. An important role of albumin in energy supply of the organism under starvation also explains the fact that the relative content of albumin in white muscle interstitial fluid increases, when the total protein decreases in the plasma and white muscle interstitial fluid in starving fishes. Thus, the protein concentration in the blood serum of mature goldfishes (females, III stage of gonad maturation) after three months of starvation (in the conditions of the keeping in aquarium) decreased 1.5 times as compared to feeding fish. In this case, the relative albumin content remained stable in serum (16.0±3.9 under feeding and 15.9±3.9 under starvation), but increased 1.47 times in white muscle interstitial fluid (5.41±0.35 under feeding and 7.90±1.33 under starvation) (Andreeva, 2008). The experimental study of other fish species (*Oncorhynchus mykiss*) also showed, that the plasma oncotic pressure, caused by the proteins, is not always the crucial factor for maintaining extracellular fluid balance in the organism (Olson et al., 2003).

Chapter 6

STRUCTURAL CONVERSIONS OF LOW-MOLECULAR BLOOD PLASMA PROTEINS DURING THE ADAPTATIONS OF PLASTIC AND WATER METABOLISM IN *TELEOSTEI*

6.1. STRUCTURAL CONVERSIONS OF FRESH-WATER BONY FISHES BLOOD PROTEINS DURING THE ADAPTATIONS OF PLASTIC METABOLISM

Structural Changes in the Low-Molecular Blood Protein Fraction During the Ovarian Cycle of Fishes

Two basic ways of organization of the low-molecular plasma protein fraction in fresh-water bony fishes - base and plastic - are described in chapter 4. The low-molecular fractions are organized according to the base type in fishes with the gonads of I-III maturity stages, and according to the plastic type in fishes with the gonads IV maturity stage. Just before the spawning the base type of the LMF organization turns into the plastic one (Figure 6.1). The disappearance of heterogeneity in the subfraction of "low-molecular proteins" and structural conversions in the subfraction of "albumins" occur in the course of such replacement. The described way of the protein transformations before spawning is found in roach *Rutilus rutilus*, bream *Abramis brama*, silver bream *Blicca bjoerkna*, muvarica *Alburnus alburnus*, zope *Abramis ballerus*, sabrefish *Pelecus cultratus*, goldfish *Carassius auratus*, pike *Esox lucius*, pike-perch *Stizostedion lucioperca*, perch *Perca fluviatilis*and other freshwater bony fishes.

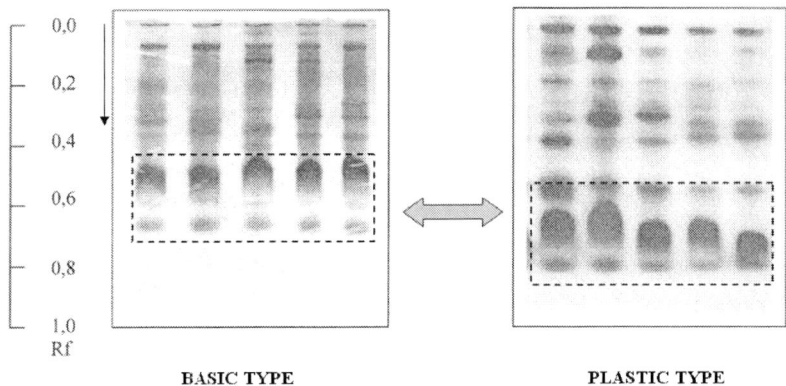

Figure 6.1. Changing of the basic type of organization of low-molecular fraction of blood plasma proteins by plastic one during ovarian dynamics of fresh water bony fishes.

In this chapter the transformations of albumins will be considered. The basic protein type is characterized by albumins with electrophoretic mobility R_f 0,55 and MM approximately 120-130 kDa under the native conditions (Figure 6.1). 10-13 subunits with MM from 13 to 73 kDa are revealed in the composition of this native albumin in SDS-PAGE(Figure 6.2); the same number of components is revealed in 2D-PAGE with 8M urea. In the both cases – in SDS-PAGE and PAGE with the urea – only one powerful macrocomponent with MM about 67 kDa (serum albumin) is distinguished for the electrophoresis spectrum (Figure 6.2).

MALDI-TOF- analysis of this protein and the search of the homologues for it in the NCBI database among all organisms didn't reveal any reliable candidate proteins. Nevertheless, the binding of albumin-specific dyes - bromphenol blue, bromcresol purple (only unspecific binding) and Evans blue - by this protein, its MM value, as well as its participation in the lipid and carbohydrate transports made it possible to consider this protein as albumin-similar one (Andreeva, 2010b).

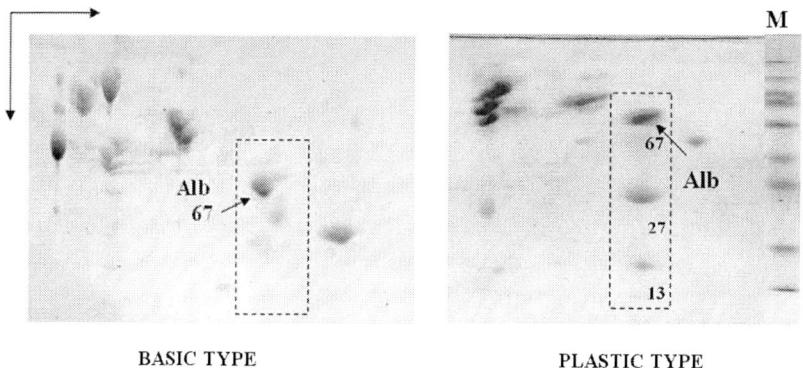

Figure 6.2. 2D-SDS-electrophoresis of serum proteins with basic and plastic types of organization of low-molecular fraction of bream and roach. Alb – serum albumin with MM about 67 kDa.The proteins in composition of oligomeric albuminare outlined by the frame. Vertical arrow shows SDS-electrophoresis direction, horizontal –disk-electrophoresis direction. M- PageRuler^{TM} Prestained Protein Ladder Plus Marker (Fermentas). (Andreeva, 2008, 2010b).

The agreement of the quantity of protein subunits in SDS-PAGE (reducing conditions) and in PAGE with 8M urea(Figure 4.11) is caused by the fact that the native protein with MM about 120 kDa is oligomer, and proteins in its composition are bound by noncovalent bonds. The dynamic of the dissociation of olygomeric protein to subunits under the action of the increasing urea concentrations obviously demonstrates shows the noncovalent nature of subunits binding in the oligomer (Figure 6.3).

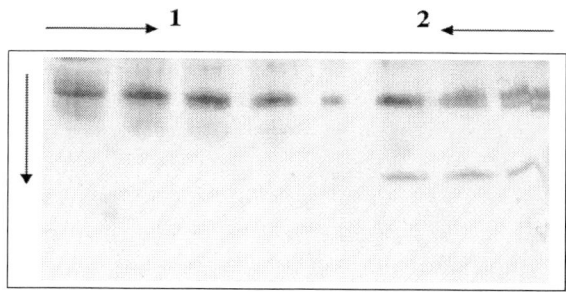

Figure 6.3. Electrophoresis of roach serum oligomeric albumin in PAGE with urea concentration gradient (0-8M). Vertical arrow show the direction of electrophoresis; horizontal arrow 1 (LTR)show urea gradient from 0 to 8M, horizontal arrow 2 (RTL) – PAGE concentration gradient from 11 to 15%. (Andreeva, 2008).

The plastic type of the organization of LM-proteins is characterized by the albumins with MM values approximately 90 kDa. The native protein is glyco- and lipoprotein. There are some subunits in its structure. Among them the macrosubunit with MM approximately 67 kDa (serum albumin) is distinguished by the high relative content in SDS-PAGE (Figure 6.2). In addition to it two subunits with MM approximately 27 and 13 kDa are revealed in the composition of olygomeric protein (in bream and roach) (Figure 6.2). By the PAS method (Gaal et al., 1982) the carbohydrates are determined in the albumin structure (Andreeva, 1999, 2010b).

The change in the subunit composition of the oligomeric protein, which contains macrosubunit with MM approximately 67 kDa (serum albumin) is caused by the active use of albumin in gonad ripening. In the process of albumin degradation the products of its proteolysis with MM 27 and 13 kDa appears in the blood flow. Probably, these products are the links in the event of albumin degradations to the amino acids, which are further used in the biosynthetic processes, which ensure the ripening of the gonads (Kirsipuu, Laugaste, 1979; Andreeva, 2010a, b).

Thus, the native albumins of basic and plastic types are oligomeric proteins, which differ by the quantitative and qualitative composition of the subunits, which enter the composition of olygomeric protein. However, the both types of native proteins contain the main macrocomponent in their structure - serum albumin with MM approximately 67 kDa. In oligomer of the basic type 10-13 different proteins are bound with the albumin molecule, and in oligomer of plastic type the products of its partial degradation are bound with the albumin molecule. Weak interactions between the subunits take part in the stabilization of oligomeric proteins; the intermolecular contacts in the oligomer are facilitated by the carbohydrates in the structure of the subunits (Andreeva, 2010b).

The conversions of the low-molecular fraction during the ovarian cycle of the marine fishes (goatfish *Mullus barbatus*, stargazer *Uranoscopus scaber*, round goby *Neogobius melanostomus*, goad goby *Mesogobius batrachocephalus*, shore rockling *Gaidropsarus mediterraneus* and corkwing *Symphodus tinca*) did not depend on the structural transformations of proteins, but they were proteins redistribution within the fraction. During gonads ripening the redistribution of the relative content of low-molecular fraction proteins occurred; the macrocomponent from the low-molecular fraction practically disappeared from the electrophoregram in scorpionfish *Scorpaena porcus* before the spawning, that indicates the active utilization of this macrocomponent (albumin) during vitellogenesis.

Structural Changes in the Low-Molecular Fraction of Blood Plasma Proteins in Roach (1+) under Conditions of High Water Temperature

The structural transformations of the low-molecular fraction of plasma proteins, described for mature fishes before spawning, are revealed in immature roach (1+) also, which were kept at high water temperature (22,5^{0}C) in the winter period in laboratory condition without the food (Andreeva, 2008). The plastic type of the organization of low-molecular proteins from such roach can be explained, probably, by the acceleration of metabolism under the effect of high water temperature. Such acceleration needs energy, and albumin is one of its suppliers. The amino acids appeared in result of the albumin degradation are used during the gluconeogenesis for the synthesis of glucose, which is the basic energy source in the starving organism.

6.2. STRUCTURAL TRANSFORMATIONS OF OLIGOMERIC PROTEIN FROM THE BLOOD PLASMA LOW-MOLECULAR FRACTION DURING THE ADAPTATIONS OF FRESH-WATER BONY FISHES TO CONDITIONS OF HIGH WATER SALINITY

The pronounced differences in the relative content of different plasma protein fractions in fishes kept in salt water 8 and 10‰ and in fresh water are not revealed. The marked changes in the blood and the interstitial fluid appeared only with the placement of fishes into the water with the salinity of 11,5‰ and higher (20‰).

The relative content of the oligomeric protein in the serum of fishes decreased almost two times at the water salinity 11,5‰, and in this case the increase of the relative content of low-molecular proteins in the serum was not marked (Figure 6.4).

A sharper reduction of the relative content of the oligomeric protein (3-4 times) occurred in the interstitial fluid of white muscles. In this case an increase in the relative content of the low-molecular proteins, including of protein with MM approximately 30 kDa, was recorded (Figure 6.4). In acute experiments with the salinity of 20‰ the cases of the complete disintegration of oligomeric protein in the tissuefluid of muscles were found (Andreeva, 2008).

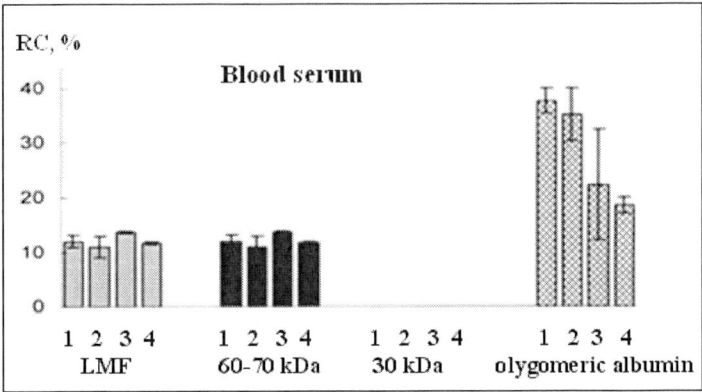
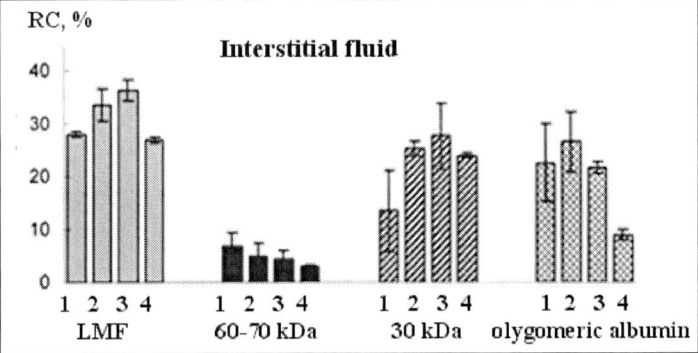

Figure 6.4. Relative content RC of low-molecular fraction LMF and its subfraction (with MMabout 60-70, 30 kDa, and olygomeric albumin) in bream serum and interstitial fluid from white muscle in fresh water (1) and salt water – 8‰ (2), 10‰ (3), 11,5‰ (4) (Andreeva, 2010b).

Thus, the increase in the water salinity led to the decrease of the relative content of oligomeric protein both in the serum and in the interstitial fluid, but the increase in the relative content of low-molecular proteins was observed only in the interstitial fluid. It allows to assume that the pool of low-molecular proteins in the interstitial fluid in the condition of high salinity is supplemented due to the dissociation of oligomeric protein in the process of its transcapillary displacement from the blood plasma into the interstitial space.

In the experiment with breams, placed into water with the salinity 20‰, the case of replacement of the oligomeric protein of the basic type by the oligomeric protein of the plastic type is registered. Such replacement occurred not in the process of the transcapillary transport of protein into the interstitial fluid, but directly in the blood serum - similarly, like what occurs during the adaptations of plastic exchange in mature fishes before spawning (p.6.1). It indicates to the possibility of structural transformations of oligomeric albumin directly in the intravascular space, but not only in the course of its displacement into the tissue fluids.

6.3. UNIVERSAL ALGORITHM OF THE STRUCTURAL CONVERSIONS OF THE PLASMA PROTEINS LOW-MOLECULAR FRACTION OF FRESH-WATER BONY FISHES

The analysis of reconstructions of the low-molecular fraction of proteins of plasma and tissue fluids in mature and immature fresh-water bony fishes under the conditions of fresh and saline (11,5‰ and 20‰) water made it possible to make some conclusions.

The main conclusion is that dynamic reconstructions of the low-molecular fraction of the blood plasma have the discrete nature. These reconstructions are caused and "facilitated" by the oligomeric organization of native albumin: about 10-13 low-molecular proteins, including the albumin with MM about 67 kDa, bound with each other with noncovalent bonds, are integrated in the oligomeric complex.

The above-described reconstructions of the low-molecular fraction during the plastic exchange in fresh-water bony fishes occur according to one scenario of replacement of the basic type of organization by the plastic one and vice versa. The detection of the case of intravascular reconstruction of plasma proteins in fishes, placed in water with the salinity of 20‰, on the same way allows to suggest, that the conversions of the low-molecular fraction of plasma protein during the adaptations of the plastic and water metabolism of the fishes *in situ* occur according to an universal algorithm, which includes the dissociation of oligomer to the subunits (to one degree or another) in the process of its transcapillary displacement into the tissue fluids. And this is the second conclusion.

And the third conclusion is that the dynamic transformations of oligomeric protein occur in the periods of adaptations of fishes to varying conditions of external and internal environment (water salinity and the intensity of metabolism) and contribute to the stabilization of the water and plastic metabolism of fishes.

Chapter 7

THE PARTICULAR ROLE OF HEMOGLOBIN AND RESISTANCE PROPERTIES OF ERITROCYTE MEMBRANES IN THE FORMATION OF FUNCTIONAL ORGANIZATION OF FISH BLOOD PROTEINS

The priority of the respiratory function of the blood determines the strategy of vertebrates for the retention of iron and all iron-comprising ligands – the products of the hemoglobin degradation. The realization of this strategy affects the functional organization of the blood plasma specialized proteins, which bind the heme (hemopexin), the iron (transferrin) and hemoglobin (haptoglobin). The special features of the realization of this strategy in fishes is considered in this chapter.

In fishes the respiratory function of the blood is mainly provided by the hemoglobin situated in the erythrocytes. The structural integrity of protein itself and continuity of erythrocyte is important firstly for the hemoglobin functioning. Meanwhile it is known that the erythrocytes of fresh water *Teleostei* are apt to the intravascular hemolysis, and hemoglobins are easily destroyed. Many factors such as stresses, pH, the temperature, the lipid composition of fodders, natural aging of erythrocytes, the glucose-6-phosphate dehydrogenases deficit in erythrocytes, the activation processes of the peroxide oxidation of lipids at alias influence the resistance of the hemoglobin and erythrocytes (Alekseev, Berliner, 1972; Idelson et al., 1975; Andreeva,

1987d, 1997, 2001a, 2006; Fainshtein et al., 1987; Kruse, Sordyl, 1988; Messager et al, 1992; Obach et al, 1993; Kiron et al, 94; Pages et al, 1995; Knoph, Thorud, 1996; Soldatov, 2005; Andreeva et al., 2006, 2009). It can be assumed that *Teleostei* should possess the special mechanisms of the respiration stabilization, which compensate the low level of the stability of erythrocytes and hemoglobin.

7.1. THE STRUCTURAL ORGANIZATION OF HEMOGLOBINS OF CARTILAGINOUS FISHES (*CHONDRICHTHYES*) AND THEIR RESISTANCE STABILITY TO DESTABILIZING FACTORS (DEHYDRATION, FREEZING)

Structure of Hemoglobins of Cartilaginous Fishes

The hemoglobins of cartilaginous fishes as hemoglobines of other fishes consist of the alpha- and beta- subunits, united in the tetrameric molecule $\alpha_2\beta_2$. In addition to the tetrameric hemoglobins there are monomeric and dimeric hemoglobins in the fishes. It is found that the genes HBA (hba) and HBB (HBB1 and HBB2) encode alpha-chains of 141-148 and beta-chains of 137-142 amino-acid residues respectively in such representatives of *Chondrichthyes* as spotless smooth hound *Mustelus griseus* (UniProtKB/Swiss-Prot: Q9YGW2, Q9YGW1) (Figure 7.1); red stingray(akaeri) *Dasyatis akajei*(UniProtKB/Swiss-Prot: P56691, P56692); spiny dogfish*Squalus acanthias*(UniProtKB/Swiss-Prot: P07408, P07409); port Jackson shark*Heterodontus portusjacksoni* (UniProtKB/Swiss-Prot: P02021, P02143); eaton's skate*Bathyraja eatonii* (UniProtKB/Swiss-Prot: P84216, P84217) (Figure 7.2). The alpha-1 and alpha-2 chains of marbled electric ray*Torpedo marmorata*consist of 141 amino-acid residues (UniProtKB/Swiss-Prot: P20244, P20245); the genes HBB1 and HBB2 encode beta-1 and beta-2 chains of 142 amino-acid residues (UniProtKB/Swiss-Prot: P20246; P20247).

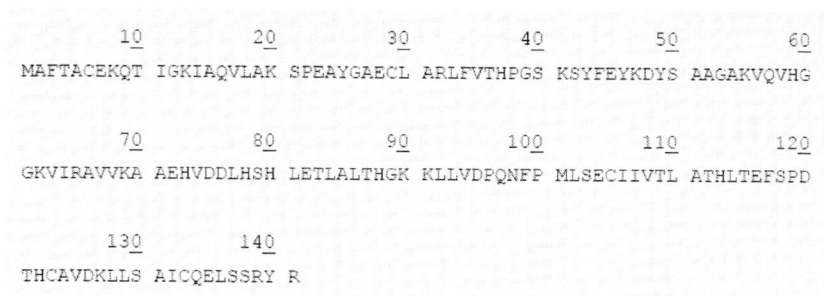

Figure 7.1. Amino acid sequence of Hemoglobin subunit alpha (complete) of Spotless smooth hound *Mustelus griseus*; Length 141, Mass 15,428 Da. UniProtKB/Swiss-Prot: Q9YGW2 (HBA_MUSGR).

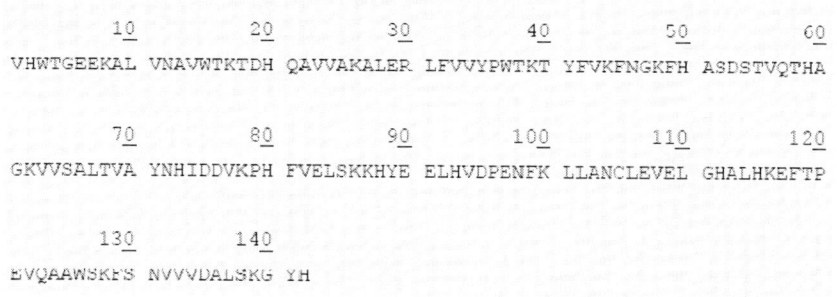

Figure 7.2. Amino acid sequence of Hemoglobin subunit beta of Squalus acanthias; Length 143, Mass 16,140 Da. UniProtKB/Swiss-Prot: PO7409 (HBB_SQUAC).

The Structural Organization of Hemoglobins of Spiny Dogfish *Squalus Acanthias* and Rays - Buckler Skate *Raja Clavata* L. and Common Stingray *Dasyatis Pastinaca* L

In the disk- electrophoresis the hemoglobins of the spiny dogfishand the buckler skateare differentiated to 6-8 components with R_f in the range 0.38-0.85, and the hemoglobins of the common stingray- to 10-12 components with R_f in the range 0.04-0.5 (Figure 7.3).

The elution profile of spiny dogfishhemoglobin from the column with Sephadex is represented by the peaks of monomer, dimer, tetrameric and does not contain of the heme of the high-molecular aggregate (Figure 7.4), which is the aggregated globin.

The protein, which does not contain the heme in its composition and has tendency toward the polymerization, was discovered in the composition of hemoglobin and in other species of *Chondrichthyes* (Nash, Tompson, 1974; Fyhn, Bolling, 1975).

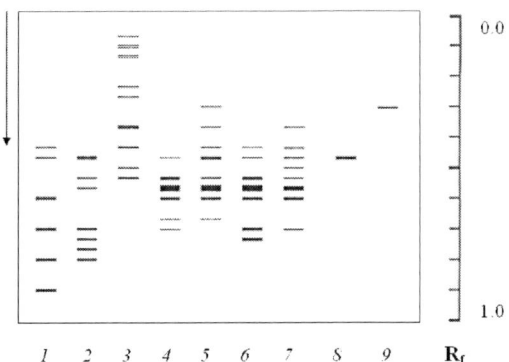

Figure 7.3. Scheme of electrophoresis of hemoglobins of some fishes: spiny dogfish *Squalus acanthias* (1), buckler skate *Raja clavata* (2), common stingray *Dasyatis pastinaca* (3), beluga *Huso huso* (4), sterlet *Acipenser ruthenus* (5), Russian sturgeon *Acipenser gueldenstaedti* (6), starred sturgeon *Acipenser stellatus* (7), pike perch *Stizostedion lucioperca* (8), bream *Abramis brama* (9). Vertical arrow show the direction of disk-electrophoresis. R_f - electrophoretic mobility. (Andreeva, 2006).

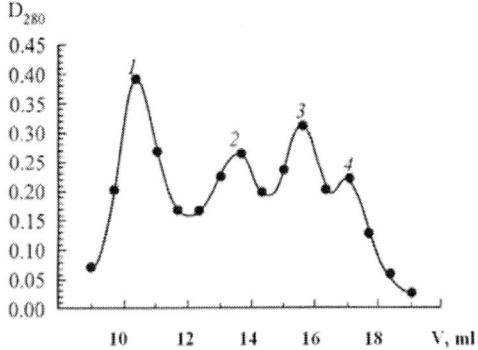

Figure 7.4. Elution profile of hemoglobin of spiny dogfish *Squalus acanthias* from column with Sephadex G-100. V – volume, milliliters. 1 – 4 - absorption peaks of high molecular aggregate (1), tetrameric hemoglobin (2), dimeric (3) and monomeric (4) hemoglobins. (Andreeva, 2006).

The hemoglobin MM values of the spiny dogfishand the rays are 68 kDa for tetrameric hemoglobin and 35 kDa for the dimeric one according to the gel chromatography data (Andreeva, 2006). The hemoglobin MM subunits of the spiny dogfish and the buckler skateare 14 kDa and of the common stingray are somewhat above - 14,5-16 kDa according to the SDS-electrophoresis.

In the process of electrophoresis under the nondenaturing conditions hemoglobin of the spiny dogfishunderwent the destruction. The introduction of 5M urea in the PAGE prevented the hemoglobin destruction and stabilized the protein in the dimeric form. The ray hemoglobins were not destroyed in the process of electrophoresis in PAGE without urea.

Hemoglobin Resistance to the Dehydration and Freezing

The dehydration of the spiny dogfishhemoglobin in the solution of ammonium sulfate occurred rapidly. The spiny dogfishhemoglobin precipitates practically instantly at all values of ammonium sulfate saturation (from 5 to 75%). All hemoglobin of the spiny dogfishturns from oxyhemoglobin into methemoglobin under the action of ammonium sulfate. Crystals of the spiny dogfishhemoglobin, grown in ammonium sulfate, were represented by the short hexagonal prisms with the angles of the base, which are repeated through two; the diameter of the crystal base is about 100 μm (Figure 7.5).

The freezing-thawing procedure led to the destruction of the spiny dogfishhemoglobin, which appeared in the disturbances of the crystal formation process (Andreeva, 1987d, 2006).

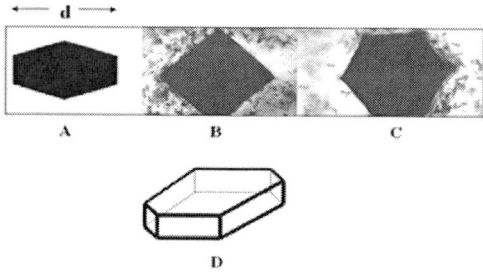

Figure 7.5. Crystals of hemoglobins in ammonium sulphate solution of spiny dogfish *Squalus acanthias* (A, D) and sterlet *Acipenser ruthenus* (B, C). d – diameter of spiny dogfishhemoglobin crystal equal 100 μm. (Andreeva, 2006).

7.2. THE STRUCTURAL ORGANIZATION OF HEMOGLOBINS FROM *ACIPENSERI FORMES* FISHES AND HEMOGLOBIN STABILITY TO THE DESTABILIZING FACTORS (DEHYDRATION, FREEZING, UREA)

Nonmigratory and Semi-Anadromous Fishes

The tetrameric hemoglobins of sterlet (*Acipenser ruthenus*) and Russian sturgeon (*Acipenser gueldenstaedti*) are differentiated to 3-4 basic and 3-5 minor fractions in the disk-electrophoresis. And each hemoglobin fraction is differentiated by several components with the identical charges and different molecular weights, which correspond to tetrameric hemoglobin, in 2D-electrophoresis (concentration PAGE gradient 3-20%) (Figure 7.6).

The hemoglobin subunit MM is approximately 16-18 kDa in SDS-PAGE. And 5M urea caused the disintegration of the tetrameric hemoglobin to the dimers.

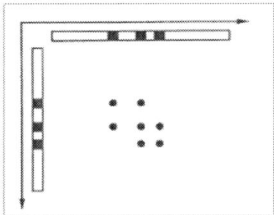

Figure 7.6. 2D-electrophoresis of sterlet hemoglobin in PAGE concentration gradient (3-40%), scheme. Vertical arrow show the direction of PAGE concentration gradient electrophoresis, horizontal arrow – of disk-electrophoresis. (Andreeva, 2006).

The hemoglobin crystals from the sterlet and Russian sturgeon, grown in the ammonium sulfate solution, were represented by small short tetra- and hexagonal prisms with the diameter "d" of base approximately 100 μm (Figure 7.5), frequently with the structural imperfections in the form of chamfered prisms, penetration twins and others.

The hemoglobins of the sterlet and the Russian sturgeon were salted out from the solutions at 20-25% saturation of ammonium sulfate. The met-form of hemoglobin predominated in the solutions of sulfate of ammonium with the saturation 5-25%, but oxyhemoglobin was also found. Only the met-form of hemoglobin was found at higher concentrations of ammonium sulfate.

The procedure of freezing-thawing led to the destruction of the sterlet and the sturgeon hemoglobin. The destruction was revealed at the disturbances of crystal formation. In the process of destructionthe hemoglobin of the sterlet and the Russian sturgeon first passed into the met-form, and then the disintegration of tetrameric hemoglobin to the dimers and the monomers occurred (Andreeva, 1987d, e; 2006).

Migratory Fishes

The hemoglobins of beluga and starred sturgeon were differentiated into 2-4 basic fractions by the charge. In contrast to the sterlet and the Russian sturgeon hemoglobins, beluga and starred sturgeon hemoglobins formed large short and long crystals in the form of tetra- and hexagonal prisms with the base diameter "d"approximately 1200 µm. They exceeded 12 times the diameter of hemoglobin crystals of the sterlet and the Russian sturgeon.

5M urea caused the disintegration of the tetrameric hemoglobin to the dimeric one.

The hemoglobins of beluga and starred sturgeon precipitate in the ammonium sulfate solution when the salt saturation was higher than 35-40%. The oxy- and deoxy-forms predominated among hemoglobins with these concentrations of ammonium sulfate.

Freezing-thawing did not lead to the destruction of hemoglobins of beluga and starred sturgeon. Even after double freezing-thawing their hemoglobins formed the regular large crystals (Andreeva, 1987d, e; 2006).

7.3. THE STRUCTURAL ORGANIZATION OF *TELEOSTEI* HEMOGLOBINS AND THEIR RESISTANCE TO DESTABILIZING FACTORS (DEHYDRATION, FREEZING, UREA)

The Hemoglobin Structure of *Teleostei*

In bony fishes the genes HBA encode alpha - chains mainly of 141-144 amino-acid residues, the genes HBB encode beta - chains of 141-148 ones (UniProtKB) (Figure 7.7, 7.8).

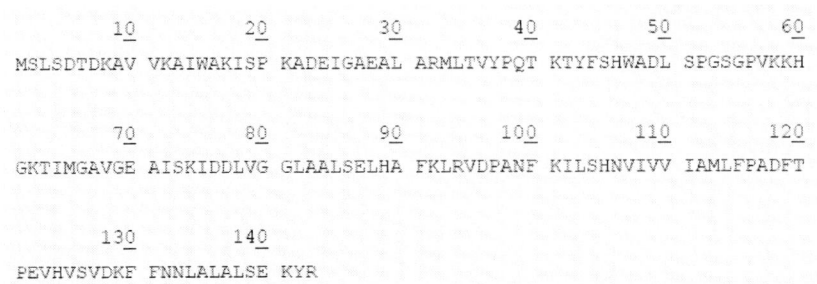

Figure 7.7. Amino acid sequence of Hemoglobin subunit alpha (complete) fromZebrafish *Danio rerio*.Length 143 AA, Mass 15,522 Da. UniProtKB/Swiss-Prot: Q90487 (HBA_DANRE).

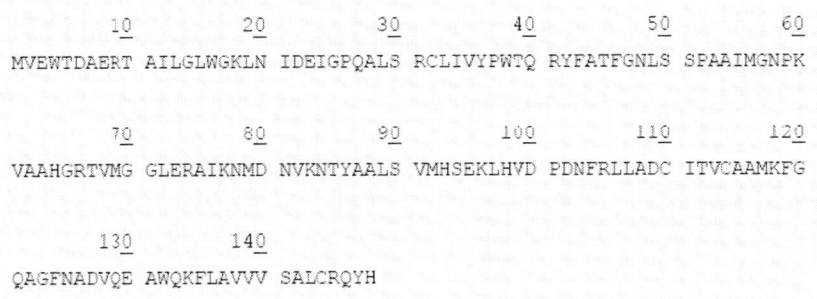

Figure 7.8. Amino acid sequence of Hemoglobin subunit beta-1(complete) fromZebrafish *Danio rerio*.Length 148 AA, Mass 16,389 Da. UniProtKB/Swiss-Prot: Q90486 (HBB1_DANRE).

In zenrafish *Danio rerio* (*Brachyodanio rerio*) the genes encoding embrional hemoglobins - hbae and hbbe - aredetected, they are located on 3 and 12th chromosomes (Figure 7.9, 7.10). Genes of adult hemoglobin are located on 3 chromosome (Figure 7.9).

Embryonic genes encode the embryonic types of alpha- and beta globin chain, for example - embryonic globin alpha a3 of 123 amino-acid residues (UniProtKB/TrEMBL: Q7T1B2). In Atlantic cod *Gadus morhua* together with alpha-chains of hemoglobin of 143 amino acids (UniProtKB/TrEMBL: B3F9J9, B3FE8 et al.) the chains of 180 (UniProtKB/TrEMBL: B3F9L9, B3F9F3) and 184 amino-acid residues (UniProtKB/TrEMBL: B3F9M2) are described.

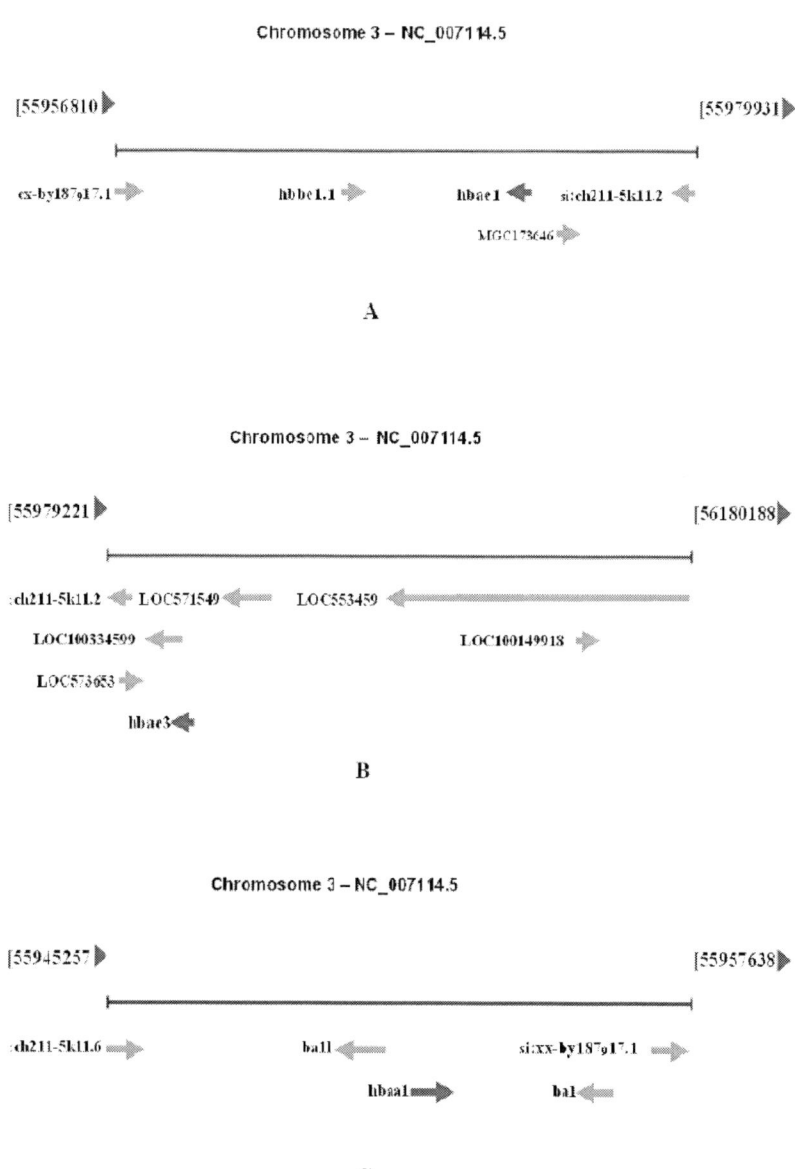

Figure 7.9. Location of embryonic (hbae1 and hbae3; A, B) and adult (hbaa1, C) hemoglobin alpha genes of Danio rerio on 3 chromosome. NCBI, Gene,RefSeq status PROVISIONAL.

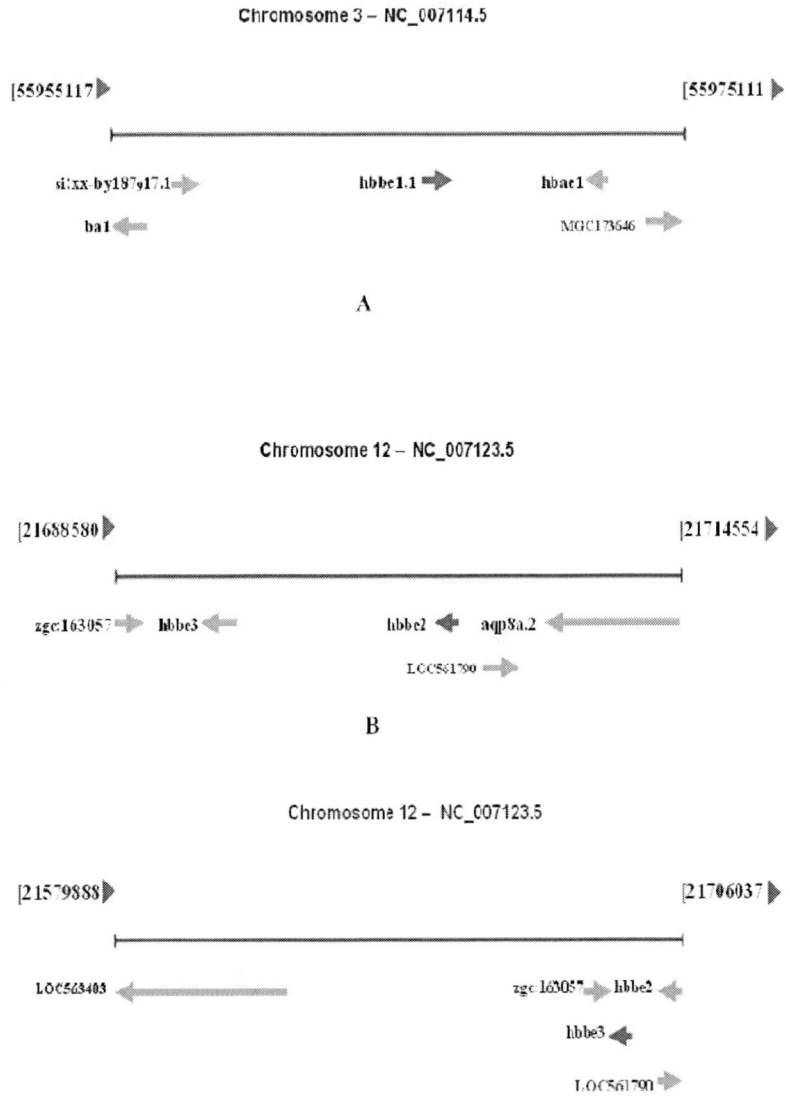

Figure 7.10. Location of embryonic hemoglobin beta genes of Danio rerio - hbbe1.1 (A), hbbe2 (B), hbbe3 (C) - on 3 and 12th chromosomes. NCBI, Gene, RefSeq status PROVISIONAL.

In Atlantic salmon *Salmo salar* the beta-1-chains of 147 (UniProtKB/TrEMBL: C0H744) and 214 amino-acid residues (UniProtKB/TrEMBL: C0H824) are described. In the same species the gene HBA4 encodes alpha-4-chain of hemoglobin of 143 amino-acid residues (UniProtKB/TrEMBL: C0H789), and the gene HBA encodes the chain of 235 amino-acid residues (UniProtKB/TrEMBL: C0H805).

The hemoglobins of bony fishes are presented by, as a rule, tetrameric forms. In the electrophoresis the minor band of the octamer is frequently encountered.

The Structural Stability of Hemoglobin of Fresh-Water Fishes

The oxy-hemoglobin of the bream, the roach, the white bream and the bleak obtained from erythrocytes washed in the physiological solution, was destroyed during 1-2 days. The destruction was accompanied by appearance of additional protein band and of free hemin on the Kohlrausch border in the disk-electrophoresis, and also by loss of peroxidase activity by the globin. Thus, the decomposition of hemoglobin in bony fishes was the result of bond breakage between the heme and the globin. In golden carp the case of spontaneous destruction inside- and extra-erythrocytic hemoglobin was revealed (Andreeva, 1997, 2006; Andreeva et al., 2006; 2009).

The Structural Stability of Hemoglobin of Marine Fishes

The hemoglobin of marine fishes was not destroyed during the year and remained predominantly in the the oxy- and deoxy-form during the storage in the distilled water above the precipitate of the erythrocyte shadows at 4^0C (Table 7.1) (Andreeva et al., 2009; Andreeva, Ryabtseva, 2011).

Stability of Bony Fish Hemoglobins to Dehydration

The hemoglobins of marine and fresh-water fishes were distinguished by the stability to the dehydration by ammonium sulfate:

1) the hemoglobins of marine fishes are least resistant to the dehydration. The hemoglobins of scorpionfish *Scorpena porcus* L., round goby *Neogobius*

melanostomus P., goad goby *Mesogobius batrachocephalus* P., shore rockling *Gaidropsarus mediterraneus* L., corkwing*Symphodus tinca* L., stargazer *Uranoscopus scaber* and goatfish *Mullus barbatus* L. precipitate at all saturation concentrations of ammonium sulfate, beginning from 5% and above (Figure 7.9). The oxyhemoglobins of these species passed into the met-form under the effect of ammonium sulfate (Andreeva, Ryabtseva, 2011);

Table 7.1. The determination of λ_{max} in the Sore range (400 – 420 nm) of the hemoglobin solution from the marine fishes every other year of his storage by 4^0C

№	The species	λ_{max} Hb, nm
1	Scorpionfish	412
2	Scorpionfish	416
3	Scorpionfish	400
5	Shore rockling	418
6	Shore rockling	416
8	Goad goby	418
9	Round goby	406
10	Golden mullet	420
12	Corkwing	412
13	Corkwing	408
14	Corkwing	418
15	Pickarel	418

(Andreeva, Ryabtseva, 2011).

2) hemoglobins of fresh-water fishes were more resistant to the dehydration: bream (*Abramis brama*) and sabrefish (*Pelecus cultratus*) hemoglobins precipitate at 60% ammonium sulfate saturation; oxy-Hb passed into the met-form of 5% ammonium sulfate saturation. Hemoglobins of zander and Volga zander precipitate at 65% ammonium sulfate saturation; at 5% salt saturation they still could be in the oxygenated form (Andreeva, 2006; Andreeva et al., 2009).

3) Pike (*Esox lucius*) and kilka (*Clupeonella cultriventris*) hemoglobins are the most resistant to the dehydration. Hemoglobin of pike can be in the oxygenated form at 70% ammonium sulfate saturation and above. On the eve of spawning the pike hemoglobin don't precipitate in ammonium sulfate solution at al. Kilka hemoglobin also remained in the oxy-form in ammonium sulfate solution, but it precipitated according to the type of bream under the effect of salt. "Two-humped" form of salting curve of kilka hemoglobin

revealed its heterogeneity by the resistance to dehydration (Figure 7.11) (Andreeva, Ryabtseva, 2011).

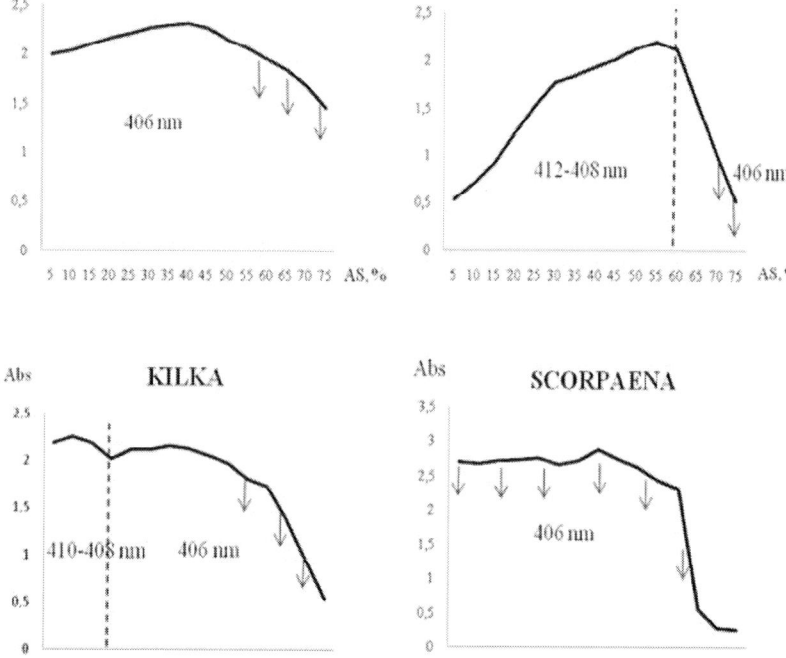

Figure 7.11. Salting out of hemoglobins of bream, pike, kilka and scorpionfish by ammonium sulphate. Abscissa axis - ammonium sulphate concentration AS (satiation, %), ordinate axis – absorption Abs in Sore γ_1- spectral range. 406 – 412 - λ_{max} of hemoglobin (nm); vertical dotted line separates oxy- and met-hemoglobins; little vertical arrows show hemoglobin precipitation at different salt concentration. (Andreeva, Ryabtseva, 2011).

Thus, the hemoglobins of kilka and fresh-water fishes are the most resistant to the dehydration, but a wide variability of this parameter was revealed among them. The hemoglobins of marine fishes precipitate under minimal concentrations of ammonium sulfate saturation and immediately change into nonworking met-form under the effect of this salt.

On the one hand, the hemoglobins of marine fishes demonstrate the high level of structural stability, maintaining the year of storage at 4^0C in the oxy-form, on the other hand – the extreme instability to the dehydration by ammonium sulfate. But meanwhile under the actual conditions proteins of namely marine fishes are subjected to the threat of dehydration.

7.4. PROBLEM OF THE INTRAVASCULAR HEMOLYSIS OF ERYTHROCYTES IN FISHES

Erythrocytes of Cartilaginous Fishes (*Chondrichthyes*)

The erythrocytes of spiny dogfish are stored two months and more at 4^0C, manifesting no signs of hemolysis. The cases of the hemolysis of erythrocytes at the sampling of blood serum and plasma are not occured as a rule.

Erythrocytes of Cartilaginous Ganoids (*Chondrostei*)

The erythrocytes of Volga sterlet, Russian sturgeon, white sturgeon (beluga) and starred sturgeon don't hemolyze during the storage also. The hemolysis of erythrocytes doesn't occur at the sampling of blood serum and plasma also.

The Effect of Dexamethasone and Testosterone as Stress Factors on the Osmotic and Acidic Resistance of Sterlet Erythrocytes
The increased content of hormones, including steroid ones, in the fish blood plasma acts as the stress factor (Selje, 1960;Armour et al., 1993; Wendelaar Bonga, 1997; Junko, Takaji, 1999; Zhou et al., 2001; Fevolden et al., 2002; Mikrjakov, 2004). This factor also effects on the resistance of erythrocytes, particularly tothe time of acidic hemolysis of the erythrocytes (Terskov, Gitelzon, 1957).

Thus, after the intraperitoneal injection to the sterlet *Acipenser ruthenus* L. (3+) of dexamethasone-phosphate (the synthetic analogs of cortisone) in the quantity, which corresponds to the level of hydrocortisone of fishes stressed (Bayunova et al., 2000), a reliable increase of the time of acidic hemolysis of the erythrocyteswas recorded in the twenty-four hours (Figure 7.12).

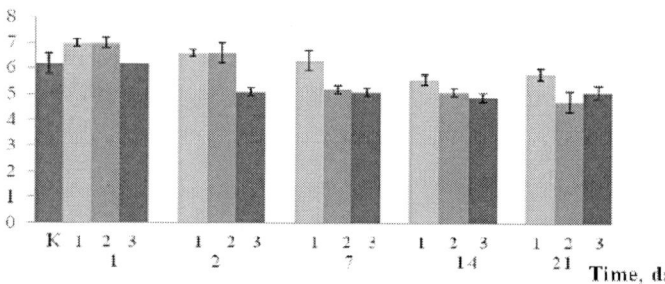

Figure 7.12. The effect of hormons (dexamethasone and testosterone) on theresistance ofsterlet erythrocytesto acid haemolysis. K- control group, 1 –fishes group,injected by physiological solution, 2 – fishes group,injected by dexamethasone, 3 – fishes group,injected by testosterone. (Ryabtseva et al., 2011).

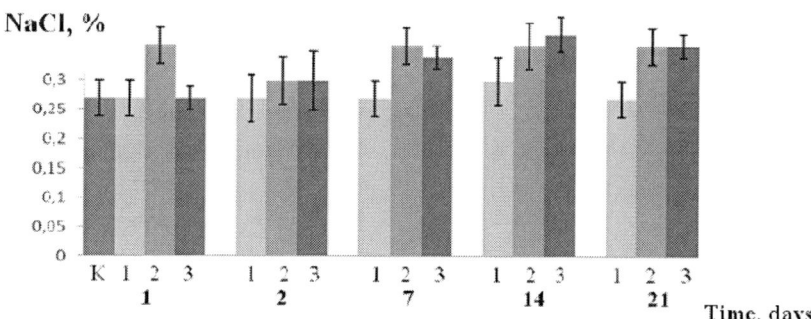

Figure 7.13. The effect of hormons (dexamethasone and testosterone) on theresistance ofsterlet erythrocytesto osmotic shock. K- control group, 1 –fishes group,injected by physiological solution, 2 – fishes group,injected by dexamethasone, 3 – fishes group,injected by testosterone.(Ryabtseva et al., 2011).

The reliable reduction in the acidic resistance of erythrocytes occurred in the 14th day. The osmotic resistance of erythrocytes under the action of dexamethasone reduced in twenty-four hours (Figure 7.13).In the fishes, injected by testosterone, the reliable reduction of acidic resistance of erythrocytes was marked on the second day of the experiment and during the subsequent days it did not change); the reduction in the osmotic resistance occurred during 14 days.In the group of the fishes, injected by physiological solution, the changes in the stability of erythrocytes are not revealed.

Thus, the action of steroid hormones as stress factors was manifested in the reliable reduction in the stability of sterlet erythrocytes to the acidic hemolysis and in the tendency to reduction in their osmotic resistance. However, no cases of the intravascular hemolysis of erythrocytes in the control and experimental groups were registered, that testifies to the high resistance indices of the membranes of the sterlet erythrocytes (Ryabtseva et al., 2011).

Special Features of the Hemolysis of Erythrocytes During Sampling and Storage of Serum and Plasma of the Blood of Mature Bony Fishes

Fresh-Water Fishes

Serum and plasma of the blood of zope (*Abramis ballerus*), bream (*Abramis brama*), bleak (*Alburnus alburnus*), silver bream (*Blicca bjoerkna*) in the spring-summer period are practically always hemolysized. In roach and sabrefish (*Pelecus cultratus*) the hemolysis of erythrocytes also occurred, but it was not permanent. In the autumn-winter period along with hemolysized serum and plasma nonhemolized ones was obtained from these fish species(Figure 7.14) (Andreeva, Ryabtseva, 2011).

The calculation of the relative content of extracellular hemoglobin in the bream and zope sera revealed the wide range of this parameter - from 0,02 to 50% and more. Meanwhile, in the pike serum the extracellular hemoglobin is practically absent, with the exception of single cases. In pike (*Esox lucius*) and zander (*Stizostedion lucioperca*) blood the erythrocyte hemolysis occurrs, as a rule, in 2-3 days since the beginning of serum settling. The tracks of the hemolysis of erythrocytes were revealed also in the serum of the goldfish (*Carassius auratus*), in which the spontaneous hemolysis of erythrocytes was revealed (Andreeva et al., 2006).

The hemolysis of bream and roach erythrocytes occurred permanently during its storage in the physiological solution at 8^0C; that is testified by the continuous increase of absorption in the range 400-420 nm (Andreeva, Ryabtseva, 2011). The bream erythrocytes began to hemolyze immediately after the blood sampling, in roach the hemolysis of erythrocytes began, as a rule, in twenty-four hours.

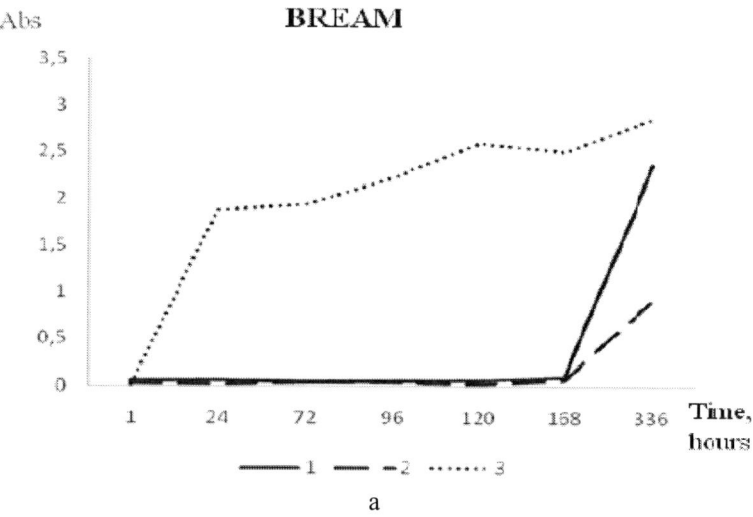

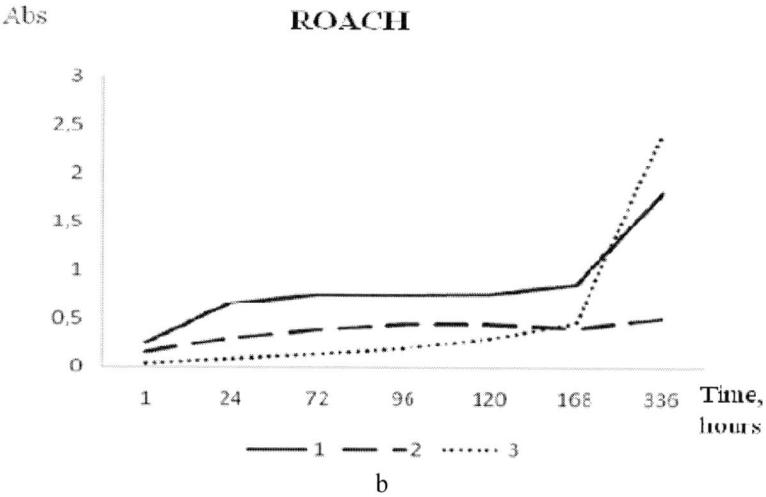

Figure 7.14. Dynamics of haemolysis of bream (A) and roach (B) erythrocytes(♀-III), when serum (1), plasma (2) and erythrocytes in physiological solution (3) were settled (1) at 8^0C. Abscissa axis – time in hours after blood sampling; ordinate axis – absorption Abs in Sore spectral range. Fishing was in autumn- winter period. (Andreeva, Ryabtseva, 2011).

Marine Fishes

During settling of blood serum in 36 example of marine fishes (the species - goatfish *Mullus barbatus*, shore rockling *Gaidropsarus mediterraneus*, round goby *Neogobius melanostomus* and goad goby *Mesogobius batrachocephalus*, scorpionfish *Scorpenaporcus*), caught in the end of May - beginning of June in Black Sea, the cases of the hemolysis of erythrocytes were not revealed at the blood sampling. Only corkwing *Symphodus tinca*erythrocytes hemolyzed immediately after blood plasma sampling.

Kilka

The kilka plasma above the erythrocytes sediment from 10 samples of the saltish water kilka (*Clupeonella cultriventris* N.), caught in August in Rybinsk Reservoir, remained nonhemolized during a week at 8^0C, after which hemolysis occurred simultaneously in all 10 samples.

Thus, the erythrocytes of saltish water (1 species) and marine (5 species) fishes don't hemolyze during settling of serum and plasma, although the exceptions occur. Meanwhile among the fresh-water fishes the species with the permanently hemolyzing erythrocytes (zope, bream) and the nonhemolyzing ones (pike) are discovered. In the majority of fresh-water fishes the erythrocytes hemolyze to one or another degree.

The data allow to assume that salinity reduction of water environment, on a whole, correlates with reduction in the erythrocytes resistance to the hemolysis of the fresh water fishes in comparison with the marine and saltish water species.

Special Features of the Hemolysis of Erythrocytes in the Hypotonic Solutions NaCl in Mature Bony Fishes

Marine Fishes

It was possible to obtain not only nonhemolized samples of the blood serum and plasma (except corkwing) practically in all sea forms, but also to preserve erythrocytes in the nonhemolized form in the process of their washing in the physiological NaCl solution. In the physiological solution the scorpionfish erythrocytes are the most resistance; they do not hemolyze during 15 minutes of the experiment. In other species (shore rockling, goatfish, stargazer, gobies) the erythrocytes hemolyzed to one or another degree after 15 minutes in the physiological solution (Figure 7.15).

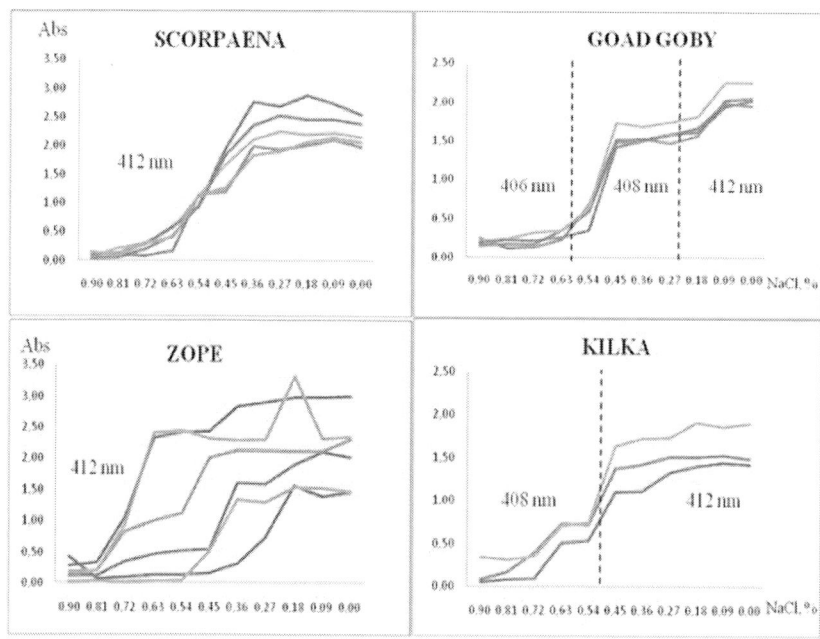

Figure 7.15. Erythrocyte osmotic resistance curves of scorpionfish (5 individuals), goad goby (4 ind.), zope (6 ind.) and kilka (3 samples to 5 ind., in all 15 ind.). Abscissa axis –NaCl concentration, %; ordinate axis – absorption **Abs** in Sore spectral range. Vertical dotted lines separateshaemolysis curves by λ_{max} (nm) of hemoglobinin Sore spectral range. (Andreeva, Ryabtseva, 2011).

The differentiation of erythrocytes at the osmotic resistance is revealed in goatfish, goad gobyand shore rockling. Thus, three fractions of the erythrocytes are discovered in shore rockling: the first in the range 0,9-0,72% NaCl erythrocytes with the methemoglobin (λ_{max} 408 nm) hemolyzed; the second in the range 0,72-0,0% NaCl - erythrocytes with met- and oxy-hemoglobin (410 nm) hemolyzed; the third fraction of young erythrocytes, did not hemolyze even in the distilled water. Under the microscope these erythrocytes in the dried and uncolored blood slide fixed by alcohol looked like round cells of an erythroid row. Old, mature and young forms of erythrocytes from goad gobyalso hemolyzed differentially: the first less resistant fraction - in the range 0,9-0,64% NaCl (406 nm), the second - in the range 0,56-0,27% NaCl (408-410 nm) and the third - in the range 0,18-0,0% NaCl (412 nm) (Figure 7.15). The hemoglobins from different fractions of erythrocytes of scorpionfish, stargazer and round goby were only in oxy-form (412-414 nm).

Fresh Water Fishes and Saltish Water Kilka
(Clupeonella Cultriventris N.) from Rybinsk Reservoir

Despite the fact that hemolysized serum and plasma are rarely encountered in pike at the blood sampling, and are frequent in bream, zope and roach, the osmotic resistance of the erythrocytes washed in the physiological NaCl solution was comparable. The erythrocytes of fresh water fishes hemolyzed or had traces of hemolysis already in the physiological solution during the first 15 minutes of the experiment.

In the pike that were caught in summer, the erythrocytes began to hemolyze in 0,9-0,81% NaCl, while in the pike, caught by autumn - in 0,72% NaCl. Mass erythrocytes hemolysis occurred in 0,5% NaCl. Moreover, in the pike, caught in summer, only oxyhemoglobin appeared in the result of the hemolysis, and in the pike, caught in autumn, first erythrocytes hemolyzed in the range 0,9-0,81% NaCl with methemoglobin (406 nm), and further in the range 0,72-0,45% NaCl - erythrocytes with met- and oxyhemoglobin (408-410 nm), and the last in the range 0,27-0,0% NaCl - erythrocytes with the oxyhemoglobin (414 nm) (Figure 7.16).

In mature roach from the reservoir the mass hemolysis of erythrocytes occurred in 0,45% NaCl, in bream and zope - it was in the range at 0,72-0,27% NaCl. In bream and roach from the ponds (age 3+ and 4+) in the spring and autumn periods the erythrocytes differed essentially by the osmotic resistance: the hemolysis of erythrocytes in autumn occurred in 0,36 % NaCl, meanwhile in spring the hemolysis began in 0,63 % NaCl. The seasonal differences can be explained not so much by the water temperature, as by the nourishment: the spring material wintered in the ponds, where the fishes did not feed more than 6 months, but autumn fishes fed actively in the finishing ponds during summer. It could not but affect the strength of the membranes of the fish erythrocytes.

The hemoglobin, which is released from the hemolysized young and ripe erythrocytes of the maturezope, bream and roach was only in the oxy-form (412-414 nm); in this case, the young erythrocytes of zope, in contrast to the bream and the roach erythrocytes, did not hemolyze even in the distilled water. The osmotic resistance of young erythrocytes of zope varied seasonally: in summer young erythrocytes did not hemolyze, in autumn they hemolyzed both in the dilute NaCl solutions and in the water.

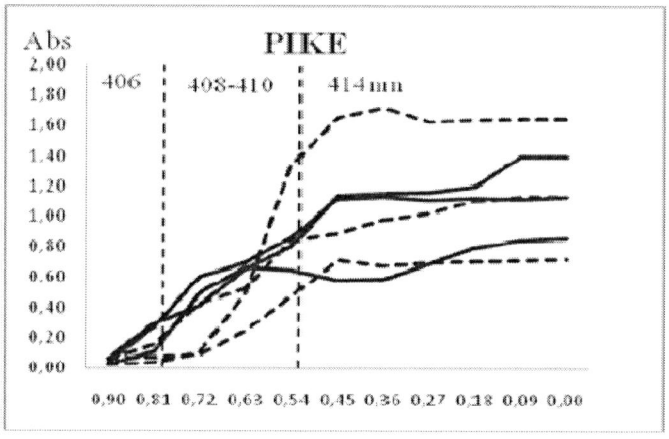

a

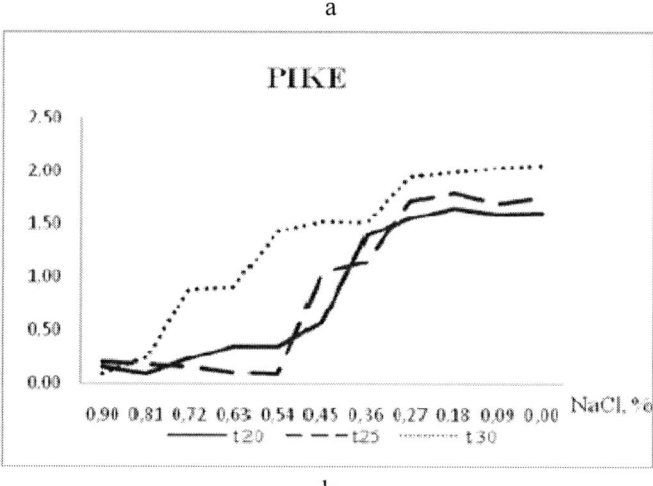

b

Figure 7.16. Erythrocyte osmotic resistance curves of the pike (A, 6 individuals) in 20°C, 25°C and 30°C (B, 3 individuals). Abscissa axis –NaCl concentration, %; ordinate axis – absorption Abs in Sore spectral range. Vertical dotted lines separates haemolysis curves by λ_{max} (nm) of hemoglobin in Sore spectral range. Pikes (A) are presented by 3 fishes, which were fished by summer (solid line) and 3 fishes, which were fished by autumn (dotted line); vertical dotted lines concern to autumn fishes. (Andreeva, Ryabtseva, 2011).

In kilka *Clupeonella cultriventris* the erythrocytes were differentiated according to their stability to the hemolysis: the unstable erythrocyte fraction began to hemolyze in 0,72% NaCl releasing methemoglobin (408 nm) out of the erythrocyte, more resistant fraction - in 0,45% NaCl, erythrocytes oxyhemoglobin (412 nm). In the redfin (*Tribolodon brandtii*) from the Sea of Japanthe first fraction of erythrocytes hemolyzed in the range 0,9-0,63% NaClreleasing the methemoglobin (406-408 nm) out of cells, the second - in 0,63-0,36% releasing met- and oxyhemoglobin (410 nm), the third - in 0,36-0,0% NaCl releasing the oxyhemoglobin (412 nm). In this case young erythrocytes of the redfin did not hemolyze in the distilled water, they were destroyed with the ultrasound; young erythrocytes contained only oxyhemoglobin.

The comparison of the curves of the hemolysis of the erythrocytes of the pike, kilka and redfin - on the one hand, and the bream, the zope and roach - on the other hand, revealed the absence of an increase of the absorption values in the initial dilutions of NaCl (0,9-0,81%) in the last group of the fishes, that can be erroneously explained by smaller stability of the erythrocytesof pike, kilka and redfin. This error is reasoned by the special features of sampling of erythrocytes for the experiment on the osmotic resistance: while the erythrocytes of the pike and the kilka did not hemolyze at the stage of erythrocytes washing by the physiological NaCl solution, the weak erythrocytes of bream, zope and roach hemolyzed at the first (of three) washing of erythrocytes.

For this reason the erythrocytes of the bream, zope and roach as the most unstable to the hemolysis were excluded from the subsequent analysis already at the stage of preparation, andonly those erythrocytes, which maintained triple washing, participated in the experiment. Therefore, the registration of the erythrocyte fractions in the experiments, which hemolyzed with the release of methemoglobin, gaves not only more complete idea about the osmotic resistance of the erythrocytic pool of the fish species, but also it made possible to draw a correct conclusion about the fact that the osmotic resistance of the erythrocytes of kilka and pike was higher than in zope, roach and bream.

The Acidic Resistance of the Erythrocytes of Fresh-Water Bony Fishes and Kilka

To obtain more adequate idea about the differentiation of the erythrocytes of fresh-water fishes and kilka according to the resistance of their membranes, the acidic hemolysis of their erythrocytes was performed, since in this procedure all existing erythrocytes of the whole blood participated (Table 7.2).

Table 7.2. Determination of the time of acidic hemolysis of fish erythrocytes

Species	Month of blood sampling	Number, sex, gonad maturity stage	Time hemolysis, sec
Kilka	September	9, females, IV	240
Pike	-"-	3	210
Pike	August	2	180
Zope	September	5, females and males, III	210
Roach	May	5, females, IV	210
Roach	September	2, males, II	180
Sabrefish	-"-	1, females, III	150
Eelpout	-"-	3, females, III	60
Whitefish	-"-	1	150
Bream	-"-	3, female and males, V and II–III	150
Bream	May	1, female, III	60
Bream	-"-	2, female and male, IV	30
Bream*	September	2	30
Zander	-"-	4, females and male, III	90
Volga Zander	-"-	3, females, III	90
Goldfish**	August	1, female, III	30

* – age 2 years 4 monthes in aquarium without food;
**– 3 monthes in aquarium, feeding(Andreeva, Ryabtseva, 2011).

It is seen from the Table that the acidic resistance of erythrocytes was reduced in the series: /kilka/-/pike, zope, roach/ - /sabrefish, whitefish, bream, eelpout/ - /zander, Volga zander/ - /the fishes, which were contained in the aquarium/. The species with the nonhemolyzed blood plasma and serum at sampling of the blood - pike and kilka - dominated according to the index of the acidic resistance of erythrocytes. However, the species with the permanently hemolyzed erythrocytes at the sampling – zope and roach were also included into the dominant group. Average parameters of the resistance of erythrocytes to the acidic hemolysis were obtained for the bream, sabrefish, eelpout, whitefish and other fresh-water species. Erythrocytes of the zander and Volga zander were the least resistant.

At the same time, keeping in the aquarian conditions of goldfish and bream both in the presence and in the absence of nourishment, led to considerable reduction in the acidic resistance of the erythrocytes. Unfavorable combination of factors probably decreased the stability of the fish erythrocyte membranes. The reduction in the resistance of erythrocytes to the hemolysis with an increase in the water temperature from 20^0C to 30^0C is an example of the effect of such factors on the erythrocyte resistance (Figure 7.16).

On a whole, erythrocytes of marine, brackish water, migratory and fresh water fishes were comparable by their osmotic resistance: the mass hemolysis of the erythrocytes of mature fishes occurred, as a rule, at 0,63-0,54% NaCl. Among studied fish species the brackish water kilka takes the special place; its erythrocytes in the composition of plasma gave no traces of hemolysis during the storage within a week. Scorpionfish (sea species) and pike (fresh water species) erythrocytes are also resistant to the hemolysis.

7.5. DECOMPOSITION OF ERYTHROCYTES AND HEMOGLOBIN IN THE GOLDFISH *CARASSIUS AURATUS* UNDER THE UNFAVORABLE COMBINATION OF THE ENVIRONMENTAL FACTORS

40 mature goldfishes of the size of 12,8-15 cm, ready to spawning, but not spawned, were placed into aquariums, where they were kept forthree months (April – July) without food at high water temperature (to $24,5^0C$) (Andreeva et al., 2006). The blood of the goldfish was brown, because its hemoglobin was in the met-form. The addition of anticoagulin and stabilizer EDTA to this blood sample led to spontaneous hemolysis of erythrocytes. The EDTA addition to the blood of other starving and feeding fishes did not provoke the hemolysis of their erythrocytes, and the hemoglobin of these fishes was in oxy- and deoxy-forms.

The electrophoresis of the total DNA from the blood of goldfish, which erythrocytes hemolyzed spontaneously, revealed the apoptotic electrophoretic spectra of DNA degradation, while the DNA of other fishes was not fragmented (Figure 7.17).

The starvation by itself is not an extreme factor for fishes, because the fishes starve during significant periods in their life cycle (Galindez, Aggio, 1998; Soldatov, 2005). But the starvation at summer temperatures, when metabolism process are intensive, is the extreme destabilizing factor. The reduction in the resistance of erythrocytes of the experimental starving goldfishes can be explained by the predominance of ripe (old) erythrocytes in their erythrocytic pool. The ripe erythrocytes accounted for about 98% of the total quantity of cells of an erythroid row in the blood smears of starving fishes, while the cells of all stages were revealed in blood smears of feeding fishes: the young oxyphilous round erythrocytes composed 10,3±1,16%; the divisible forms composed 3,8±0,82% and the ripe erythrocytes - 85,9±1,23% (Andreeva et al., 2006). The goldfish with apoptotic DNA did not differ from the other feeding and healthy fishes in the motor activity and behavior. It can be assumed that the apoptosis could be the consequence of the rearrangement of organism homeostasis from the regime of permanent erythropoiesis to the discrete erythropoiesis and apoptosis for the adaptation of fish to the extreme conditions.

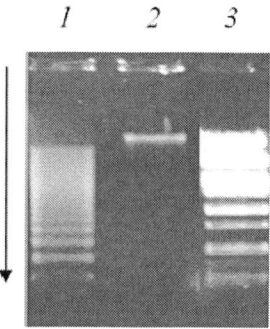

Figure 7.17. Electrophoresis of DNA from blood of goldfish *Carassius auratus* in agarose gel: 1 -DNA from starving goldfish (aquarium condition); 2 - DNA from feeding goldfish (natural reservoir); 3 - Lambda DNA/Pstl Marker. Vertical arrow show the direction of electrophoresis. (Andreeva, 2008).

7.6. FORMATION OF ERYTHROCYTES OSMOTIC RESISTANCE AND HEMOGLOBIN STABILITY TO DEHYDRATION IN BREAM AND ROACH ONTOGENESIS

The underyearling erythrocytes of bream and roach are the most resistant to the hemolysis, including acidic, in comparison with adult fishes. The osmotic resistance of erythrocytes reached minimum values in the third year and it did not change further in the ontogenesis (Figure 7.18) (Andreeva, Ryabtseva, 2011).

The stability of hemoglobin to the dehydration was minimal in underyearlings; it reached the maximal values in the third year of life and further it did not change in the bream and roachontogenesis (Figure 7.19) (Andreeva, Ryabtseva, 2011).

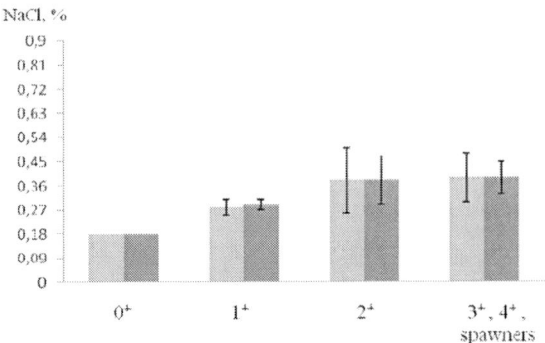

Figure 7.18. Osmotic resistanceof erythrocytesofroach (light key) andbream (dark key) ofdifferent age (0+ - 4+ and spawners). Ordinate axis – concentration ofNaCl, %. (Andreeva, Ryabtseva, 2011).

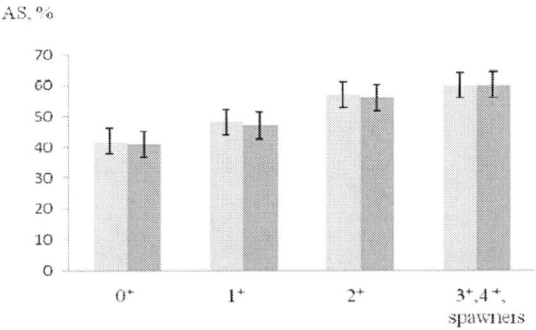

Figure 7.19. Change ofcritical concentration of precipitation of hemoglobin of roach (light key) and bream (dark key) ofdifferent age (0+ - 4+ and spawners). Ordinate axis – concentration of saturation of ammonium sulphate AS, %.(Andreeva, Ryabtseva, 2011).

7.7. THE BARRIER ROLE OF ERYTHROCYTE MEMBRANE ON THE WAY OF THE ENTRY OF HEMOGLOBIN FROM ERYTHROCYTE INTO THE BLOOD STREAM

The study of the structural organization and parameters of the stability of hemoglobin (located out of the erythrocyte *in vitro*) revealed its susceptibility to destruction processes under the action of different factors, which simulate the unfavorable actions of the environment (the dehydration under the action of ammonium sulfate, freezing, the influence of urea, the storage). The hemoglobin destruction is manifested as the conversion of protein from the oxy- and deoxy- into nonworking met-form; in the form of tetrameric disintegration to the dimers and the monomers, the hemin and the globin; in the form of the precipitation of hemoglobin with the dehydration. The hemoglobin of bony fresh water fishes with the erythrocytes, apt to the intravascular hemolysis, began to destroy to the heme, the globin and Fe^{3+} after entry into the blood flow during 1-3 days. The membrane of erythrocyte is the only barrier on the way of hemoglobin into the extra-erythrocytic space. What mechanisms in fishes prevent hemoglobin ejections into the blood flow and make it possible to preserve working oxyhemoglobin inside the erythrocyte?

1) The mechanism of the compensating type in the bream and roach ontogenesis, when the high stability of the underyearling erythrocytes compensated the instability of underyearling hemoglobin to the dehydration. The osmotic resistance of erythrocytes reduced during development of fishes and reached the minimal values in the third year of life and it did not change further; while the stability of hemoglobins increased and reached stable values in the third year of life;

2) The mechanism of the formation of differentiated stability of young and ripe erythrocytes to the hemolysis, described for zope, redeye and shore rockling, when old erythrocytes, which containing nonworking methemoglobin, destroyed first of all, and young erythrocytes, containing oxyhemoglobin, were resistant to the hemolysis;

3) The mechanism of the stabilization of the blood respiratory function by the re-posting the homeostasis under the extreme conditions, by the example of the goldfish, when in the discrete regime the replacement of all erythrocytes and hemoglobin occurs.

On a whole the erythrocytes of marine and fresh water fishes are comparable by their osmotic resistance. However, the tendency toward the intravascular hemolysis of erythrocytes is revealed only in the group of fresh-water fishes. The significant fluctuations of the level of the mineralization of fresh waters and salt composition of the internal liquid medium of the organism can contribute to such property of erythrocytes (Martemyanov, 2001; Andreeva, 2006). Along with, the wider variability of the parameters of the resistance of erythrocytes (to the hemolysis) and structural stability of hemoglobin is marked in fresh water fishes in comparison with the sea forms: the erythrocytes and hemoglobin are more durable in active and predatory species (pike, zander, Volga zander), the cases of sampling of the hemolysized blood in them are encountered rarely; while both erythrocytes and hemoglobin are less stable in less mobile not predatory forms, but the hemolysis has often permanent nature (zope, bream, roach). In other words, differences in the resistance of erythrocytes and hemoglobin in fresh-water fishes correlate with the different ecological conditions, and differ in predatory pelagophile (zander), benthophage and mesobenthophile (bream), predators - ambuscader (pike) and at alias.

It is more difficult to explain the fact that the stability of hemoglobins to the dehydration in fresh-water fishes is considerably higher than in marine fishes. On the one hand, namely marine fishes but not fresh water fishes face the challenge of the real threat of dehydration. On the other hand, aging of erythrocytes results in cardinal changes in their metabolism with the simultaneous dehydration of cytoplasm cell (Serpunin, Likhatchyova, 1998; Phillips et al, 2000). Low resistance to the dehydration of hemoglobins of marine fishes can be explained by the fact that their erythrocytes are not apt to the intravascular hemolysis, but the resistance properties of their membranes ensure the necessary autonomy of intra-erythrocytic space for fragile hemoglobins. However, the stability to the dehydration of hemoglobins of fresh water fishes, probably, is necessary in order to resist the abovementioned factors - wide fluctuations of salinity of both the fresh waters and the internal fluids of organism. Such instability of the salinity of external and internal media can be overcame due to the differential resistance of erythrocytes (zope, roach, shore rockling) and differential stability of hemoglobins to the dehydration (kilka).

7.8. STRATEGY OF THE STABILIZATION OF THE INTERNAL FLUID ENVIRONMENT OF THE FISH ORGANISMS WITH THE HELP OF THE BINDING OF THE LIGANDS CONTAINING IRON BY THE BLOOD PLASMA PROTEINS

Specialized Proteins of Cartilaginous and Bony Fishes, which Bind the Ligands Containing Iron

The structural organization of these proteins is considered in Chapter 2.

Among specialized proteins, which bind the ligands containing iron, only transferrins are discovered in all taxa of *Pisces*. The another protein -

hemopexin, which binds hemin, is discovered in cartilaginous and bony fishes, but is not found in *Acipenseriformes*. The specialized protein haptoglobin, which binds intravascular hemoglobin, is discovered only in bony fishes, but is not found neither in *Selachii* nor in *Acipenseriformes*.

The Binding of the Ligands Containing Iron by the Unspecialized Proteins of the Fish Blood Plasma by the Example of Bream

In addition to the specialized proteins, such as transferrin, hemopexin and haptoglobin, the unspecialized proteins of fresh water bony fishes can also bind the ligands containing iron (Figure 7.20) (Andreeva, 2001a).

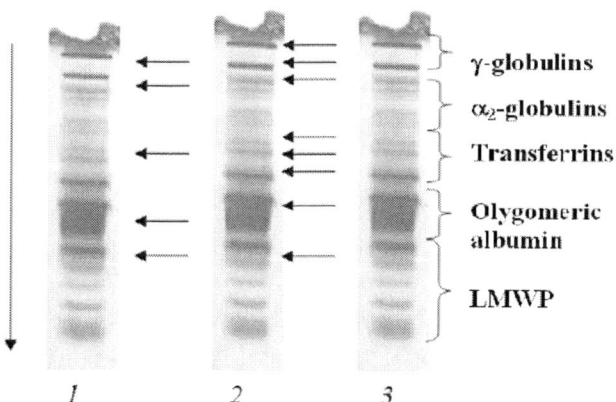

Figure 7.20. Disk-electrophoresis of bream blood serum proteins: *1* – nonhemolized serum, *2* - hemolysized serum. Vertical arrow show the direction of electrophoresis, little pointers indicates to proteins, which bind iron-containing ligands, from γ-, α_2-, β- (transferrins) and olygomeric albumin and low-molecular weigth protein (LMWP) fractions (*3*). (Andreeva, 2001a).

Such unspecialized proteins which bind ligands and contain iron are absent in the group of cartilaginous fishes - in spiny dogfishand and in the rays stingray and buckler skate, and in *Aciperiformes* fishes. The binding of hemin by serum proteins is not revealed in sterlet by one author (Andreeva, 2001a), other authors revealed the binding of hemin by albumin (Chikhachev, 1982). And only in bony fishes, or more precisely - in the group of fresh water bony fishes - the binding of the ligands containing iron by practically all proteins of the blood plasma is discovered (Andreeva, 2001a).

Among serum proteins of bony fishes, binding unspecifically iron containing ligands, there are 1) gamma - globulins, 2) alpha-1- globulins, 3) albumins and 4) low-molecular proteins.

In the nonhemolized blood serum samples from 10 breams not more than 1-3 components was stained by Mueller reagent - nitroso-R-salt (Palmour, Sutton, 1971) to Fe3+ simultaneously. Gamma-globulins, alpha-2- globulin (haptoglobin), beta- globulins (transferrin), albumin and low-molecular proteins are among them. In the blood serum samples hemolysized the Mueller reagent stains simultaneously 6-9 components instead of 1-3, among which additional proteins in the zones of mobility of gamma-globulins and transferrins are discovered. The incubation of the blood serum nonhemolized with the surplus of the salt containing iron led to the stable binding of Fe^{3+} not only by transferrin, but by albumin also. It was possible to register the binding of iron by low-molecular proteins in two samples from 20.

The incubation of the hemolysized and nonhemolized sera of the bream with hemoglobin and hemin made it possible to reveal the binding of ligands by gamma-2- and gamma-1-globulins, alpha-2- globulins (haptoglobin) and albumins.

The binding of the ligands containing iron by the unspecialized proteins – gamma-globulins, alpha-1-globulins, albumins and low-molecular proteins – occurs randomly (Andreeva, 2001a).

The attention should be paid to the fact that among the unspecialized proteins only albumins are capable to binding all forms of the ligands containing iron - Fe^{3+}, hemin and hemoglobin. The studies of the time dynamics of the hemoglobin degradation and of Fe^{3+} -, hemin- and hemoglobin-binding function of the bream blood serum proteins allow to propose the following sequence of events: the native hemoglobin entering the blood flow in result of the intravascular erythrocyte hemolysis is bound covalently by haptoglobin and noncovalently by gamma-2-globulins. In twenty-four hours the bonds between heme and globin in hemoglobin bound by haptoglobin are destroyed. The albumin binds the extracellular hemoglobin too. In contrast to haptoglobin it can bind both native and partially destructured hemoglobins.The hemoglobin bound by albumin undergoes further destruction with the release of hemin. The hemin, which is released in the result of the degradation of free hemoglobin (in the blood flow), and of hemoglobin, bound by haptoglobin and albumin, is bound further by hemopexin and by the hemin-binding centers of oligomeric albumin (Andreeva, 2001a).

Thus, the oligomeric protein albumin from the blood low-molecular fraction is the typical polyfunctional protein, which "duplicates" successfully the functions of the specialized proteins - transferrin, haptoglobin and hemopexin.

7.9. THE ROLE OF THE LOW RESISTANCE OF ERYTHROCYTE MEMBRANE AND STRUCTURAL INSTABILITY OF HEMOGLOBIN FROM FRESH-WATER BONY FISHES IN THE DECREASE IN THE LEVEL OF THE SPECIALIZATION OF THE BLOOD PLASMA PROTEINS OF FRESH WATER *TELEOSTEI*

According to the quantity of proteins, which bind the physiologically significant ligands containing iron, all *Pisces* can be subdivided into two groups: I - this group includes the *Selachii* and *Acipenceriformes*, in which no more than two blood serum proteins, which bind Fe^{3+} and hemin (transferrin, hemopexin) are discovered; and II – this group includes the fresh-water bony fishes; about 9 and more proteins, which bind the Fe^{3+}- containing ligands, are discovered in this group; among them there are specialized proteins (transferrin, hemopexin, haptoglobin) and unspecialized ones (practically all proteins of the blood plasma).

In this case, in *Selachii*, sturgeon fishes and marine bony fishes the absence of the intravascular hemolysis of erythrocytes is noted, and in many species of fresh-water bony fishes the intravascular hemolysis have the permanent nature, which leads to a constant ejection of free hemoglobin into the blood, further such hemoglobin undergoes the destruction to hemin and iron in the blood flow.The low level of the erythrocytes resistance and the structural instability of hemoglobin in fresh water bony fishes were manifested as mass decomposition of erythrocytes and hemoglobin in goldfish under the effect of unfavorable factors. However, this destruction was not fatal for the fishes. The special system "of insurance" contributes to the survival of fishes in such critical situations in the species, which have rapidly degrading hemoglobin and erythrocytes inclined to the intravascular hemolysis, the system is built of the operational binding of both hemoglobin and products of its degradation not only by the specialized proteins (by haptoglobin, hemopexin and transferrin), but also by all proteins of the plasma. This binding system is sufficiently capacious, because it maintains the load in the form of massive emission of methemoglobin from spontaneously hemolyzing erythrocytes as a result of apoptosis. The binding of the ligands containing iron by the blood plasma proteins of fresh water bony fishes is the protective mechanism, which stabilizes not only the respiratory function of the blood, but also, as a whole, the internal fluid environment of organism, protecting and "cleaning" it from the fragments of destroyed hemoglobin.

Thus, an increase in the number of blood plasma proteins, which bind the hemoglobin and products of its destruction in fresh-water bony fishes is caused by physiological strategy of organism for averting of the iron losses. The decrease in the level of the specialization of the blood plasma proteins of fresh-water *Teleostei* is the results of this strategy.

Chapter 8

PRINCIPLES OF THE FISH BLOOD PROTEIN ORGANIZATION

In spite of fundamental similarity in the organization of the blood proteins of mammals and fishes, the fish proteins have some essential features. In essence, these special features concern of low-molecular fractions, but not specialized globulins. The conservatism in the organization of globulins can be explained by the consequence of the retention of the specialized functions of the binding of the ligands containing iron by these proteins by selection.

8.1. THE FEATURES OF THE STRUCTURAL ORGANIZATION OF THE BLOOD PROTEINS OF *PISCES*

In spite of the variety of the blood plasma proteins in cartilaginous and bony fishes all proteins have the general principles of organization.

Presence of Carbohydrate in the Structure of Proteins

Not all proteins of the mammalian blood plasma are glycoproteins: only globulins contain the carbohydrate in the structure of molecule. In many fishes all blood plasma proteins - globulins, albumins (albumin-similar proteins) and low-molecular proteins - are glycoproteins. The carbohydrates are not discovered in the structure of serum albumins in *Acipenserifirmes* and *Salmonidae* fishes. Although bulltrout (*Salmo trutta trutta*) serum albumin, in contrast to other *Salmonidae*, contains sialic acids (Metcalf et al, 1998a, b). The proteins of fishes and mammals can be essentially distinguished by the strength of the carbohydrate -protein bonds in glycoprotein (Valenta et al., 1976; Andreeva, 1987b; Kirpichnikov, 1987).

The result of the glycosylation of the fish proteins can be 1) the increase of the heterogeneity level of plasma proteins, 2) the complication of the surface structure of proteins, 3) the increase of the size of the protein molecule, 4) strengthening of intermolecular interactions of proteins, which are manifested in their ability to aggregating, the appearance of protein complexes and oligomeric proteins. To the greatest degree these features appeared in the proteins of fresh-water bony fishes.

The Plurality of the Low-Molecular Proteins of the Blood of Fishes

The plurality of the proteins of low-molecular fractions is the most expresed in cartilaginous and bony fishes. Their low-molecular fractions contain up to 10 and more proteins, including albumins or albumin-similar proteins. *Acipenserifirmes* fishes differ sharply from cartilaginous and bony fishes. Their (*Acipenserifirmes*) low-molecular fractions are presented exclusively by albumins, which plurality on the electrophoregram is low and is determined by the quantity of alloforms, that do not exceed three bands in diploid forms, and six bands in polyploid ones (Russian, Siberian and Amur sturgeon).

Different Methods of Organization of the Low-Molecular Fish Blood Proteins

The low-molecular fraction in cartilaginous fishes is formed due to oligomerization or covalent modification of one or two low-molecular proteins; in *Acipenserifirmes* fishes it is represented by different albumin allovariants; in bony fresh-water fishes the low-molecular fraction consists of two discrete subfractions, which consist of the nonhomologous proteins, which are in concord reconstructed at changes of the plastic and water metabolism.

Complication of the Structural Organization of the Blood Proteins in the *Pisces*Evolution

The analysis of the quantity of blood proteins of the fishes on the 2D-electrophoregrams showed that at the similar degree of the differentiation of native proteins in PAGE (20-28 components), the quantity of subunits, which form protein variety, is substantially distinguished in different *Pisces*taxa (Table 8.1).

Thus, the variety of the blood proteins in cartilaginous fishes is ensured by the small number of subunits (13 and 17 subunits). In cartilaginous ganoids the number of subunits is more three times (46 subunits), while in fresh-water bony fishes - 7,5 times (98 subunits) in comparison with spiny dogfish *Squalus acanthias* (13 subunits). Thus, the proteins of fresh-water *Teleostei* are organized most complicatedly.

Table 8.1. The maximal quantity of proteins in PAGE at the fractionation of native and denaturative fish blood serum proteins in 2D- electrophoresis*

Taxon	The quantity PAGE components (native conditions)	The quantity PAGE components (denaturing conditions)	
		8M urea	SDS
Chondrichthyes :			
Shark(1)	28		13
Ray(2)	20	10	17
Chondrostei :			
Nonmigratory(1)	24	26	46
Semi-anadromous(1)	24	26	46
Migratory(2)	28	25	41
Teleostei:			
Salt-water (18)	26	17	43
Fresh-water(20)	26	54	98

*The number of fish species is indicated in the round brackets.(Andreeva, 2010b).

The increase in the number of subunits is ensured by the the proteins consisting of subunits; among them immunoglobulins, haptoglobins and oligomeric protein- albumin: the immunoglobulins consist of two types of subunits (H- and L- chains), haptoglobins - of two types of subunits also (alpha- and beta-chains), and oligomeric albumin - of 10-13 subunits. Taking into account the presence of immunoglobulins and haptoglobins in both groups of fishes, the two-fold increase in the number of subunits in fresh water bony fishes (98) as compared to marine ones (43) is providedexceptionally by oligomeric protein- albumins.

The Effect of the Composition of the Fish Extracellular Fluids on the Structural Organization of Blood Plasma Proteins

The basic osmotically active components in the fish extracellular fluids are salt, urea, trimethylaminoxide TMAO and proteins (Table 8.2).

Table 8.2. The concentration of osmotically active substances in the blood plasma of Chondrichthyes and Osteichthyes[*]

Fishes	Concentration, %			
	Salt	urea	TMAO	protein
Chondrichthyes sea-water fresh-water	1.42-1.77	1.14-3.6 0.49-0.78	0.53-1.5	0.4-4.3
Acipenseriformes (sea- and fresh-water)	1.7	0.02	<0.53	2.8-6.5
Teleostei	1.16-1.28	0.02-0.03	0.05	2.0-6.5

[*] Stroganov, 1961; Shilov, 1985; Schmidt-Nielsen, 1979; Anderson et al., 2002; PillansFranklin, 2004; Speers-Roesch, Ballantyne, 2006; Villalobos, Renfro, 2007.

In different taxa of fishes the osmoticity of internal fluids is supported by the different content and the relationship of these substances, in consequence of which the internal fluids of cartilaginous fishes are hypertonic, and of bony fishes - hypotonic in respect of the sea water. The hypertonicityof the internal fluids of cartilaginous fishes is maintained mainly by urea, but not by salts, which are two or three times less in Selachii than in the sea water. The internal fluids in *Acipenseriformes* and bony fishes (as fresh-water, so sea-water and migratory species) are hypotonic in respect to the sea water. Their tissue fluids contain approximately so many salts as the tissue fluids of Selachii, and ureas - almost 100 times less; probably, because of this, in the extracellular fluids of bony fishes exceeds two – five times the content of proteins in cartilaginous fishes.

How do the salt and the urea in the composition of the extracellular fluids effect on the structural organization of the fish proteins? It is known that salts, NaCl for example, in the 2M concentration and higher, are the cause of the dissociation of oligomeric proteins to the monomers. However, the salts content in the fish fluids does not exceed 1,77%, which corresponds to the concentration of 0.3M NaCl. Such salt concentration does not cause the dissociation of oligomers (Andreeva, 1997). As regards of the urea, it is established that the dissociations of the blood plasma oligomeric proteins occur at urea concentration about 3% (0.5M). The observation of the effect of different urea concentrations (urea concentration gradient from 0 to 8M) on the roach blood proteins revealed the dissociation of oligomeric protein at the urea concentration about 1M (Chapter 6).

Thus, in cartilaginous and bony fishes the structural organization of proteins under the conditions of organism is adapted to different in composition biological extracellular fluids, which contain the different urea concentrations. The proteins of the cartilaginous fishes in the absence of urea (*in vitro*)are the aggregated forms; the introduction of the urea into the reaction medium leads to the dissociation of aggregates to the low-molecular proteins and reduction in the total quantity of proteins on the electrophoregram(such effect is caused by aggregation of small proteinsof the sometypes with identical MM values in aggregates with different MM values). The aggregation of alpha-1- globulins and simultaneously - the dissociation of oligomeric immunoglobulins are found in sturgeon fishes under the effect of urea. The cases of aggregation of the blood plasma proteins under the effect of urea are found in some sea *Teleostei* (bank cod *Gadus morhua*, arctic flounder *Liopsetta glacialis*). And only in fresh-water bony fishes the oligomeric complexes dissociate into the proteins under the effect of urea, that leads to the increase of the total quantity of proteins on the electrophoregram about two times - from 26 to 54 components. Thus, the high urea concentrations in the organism internal fluids of cartilaginous fishes prevent the aggregation of proteins. Meanwhile the insignificant quantities of urea in the blood of sea *Teleostei* also support the unaggregated forms of low-molecular proteins, but in the blood of fresh-water *Teleostei* they support precisely aggregated forms of several proteins of low-molecular fraction, which enter the composition of oligomeric albumin.

The different strategies in *Chondrichthyes* and *Osteichthyes* - to the the hyperosmotic and hyposmotic internal fluids, and the different composition of these fluids in cartilaginous and bony marine and fresh-water fishes also, determined special mode of organization the proteins in different taxonomic groups and biotopes of fishes.

The Influence of the Water Salinity on the Organization of the Blood Plasma Proteins of the Fishes

The change in the salinity of water leads to the different methods of the reorganization of low-molecular fractions in the different taxonomic groups of *Osteichthyes*.

Sterlet

An increase in the permeability of the capillary walls to albumins occurs in fresh-water nonmigratory species (starlet) during its adaptation to the salinity 20‰ under the conditions of experiment; the alternation of the sea period to the river one (and vice versa) in migratory fishes leads to the change in the frequencies of different albumin allovariants.

Scorpionfish

The desalination of sea water under the experimental conditions leads to change in the relative content of the separate low-molecular plasma proteins of scorpionfish (*Scorpaena porcus*) and redistribution of these proteins within the framework of fraction, but not to the structural transformations of proteins themselves.

Roach, Bream

The dynamic conversions of the low-molecular fraction in the form of the dissociation of oligomers to the monomeric proteins and the redistributions of the pool of the low-molecular proteins between the intravascular and extravascular spaces occur in these fresh-water bony fishes under the increase of the water salinity. Namely, this capability of the blood plasma proteins of fresh-water bony fishes to dynamic reconstructions under the changes in the water salinity allows to distinguish fresh-water *Teleostei* among all *Pisces*. The detection of this mechanism made it possible to consider the salinity factor as one of the factors forming the structural variety of albumins according to the type monomer/oligomer.

Models of the Structural Organization of the Fish Blood Proteins

The ideas about the structural organization of the blood plasma proteins and their trans-capillary exchange in mammals are based on the model of the large monomeic proteins, capable to penetrate through the wall of capillary to the interstitial space in some divisions of capillary network.

We propose several models of proteins for the fishes, by the reason of large differences in their organization in different taxa of *Pisces* (*Chondrichthyes, Chondrostei, Teleostei*) and biotopes (sea and fresh-water):

- the proteins in cartilaginous fishes are organized according to the type of the monomeric proteins in the environment with the urea; in the environment without the urea the proteins are represented by aggregates. The internal fluid of the environment of cartilaginous fishes contains high concentration of urea, therefore the plasma proteins are presented as unaggregated forms in such medium. Plasma proteins of cartilaginous fishes are filtered in the interstitial space in all divisions of capillary network;
- in cartilaginous ganoids the plasma proteins are organized in the form of monomers, but oligomeric immunoglobulins are also found *in vitro*. The blood plasma proteins are filtered into the tissue space in all divisions of capillary network, the permeability of the walls of capillaries to the different proteins have the selective nature;
- in marine bony fishes the plasma proteins are organized according to the mobomeric type. Heterogeneous low-molecular fraction consists predominantly of monomeric proteins, but some species have the oligomeric albumin in the composition of this fraction;
- in fresh-water bony fishes the plasma proteins are monomers and oligomers. The latter are encountered among the serum immunoglobulins and in the low-molecular fraction. The oligomeric albumin has different subunit composition at different stages of plastic metabolism. Two discrete structural types of oligomeric albumin are revealed, which are distinguished by the quantitative and qualitative of composition subunits. The oligomeric albumin can dissociate to the composing its low-molecular proteins during the trans-capillary

exchange. The blood plasma proteins are filtered into the tissue space in all divisions of the capillary network, the permeability of the walls of capillaries to the different proteins have the selective nature. Oligomeric proteins in the composition of low-molecular fraction are detected in saltish-water and migratory bony fishes also. In other words, oligomeric proteins are revealed only in the blood of those fishes, that have (or had in evolution) fresh-water phase in their life cycle.

Similarity and Difference of Albumin Organization in *Pisces* and *Mammalia*

In the different structural models of the blood plasma proteins the basic differences concern the organization of the low-molecular fraction, including of albumins. The weak antigenic identity of the albumin-similar proteins of fishes from different taxa confirms serious differences in their surface structures (Zorin and other, 1994) and primary structures. The comparison of the amino-acid sequences of albumins in the Atlantic salmon*Salmo salar*, human*Homo sapiens* and others mammals has not revealed identical motives (DB SwissProt) (Figure 8.1); and there are no identical peptides among fragments of tryptic cleavage of the fish and mammalian albumins.

However, in some fish groups, namely - in *Acipenseriformes* and sarcopterygiian fish (*Dipnoi*), - the motives in the structure of albumin genes, similar with the motives of the mammalian albumin genes are discovered (Chikhachev, 1982; Filosa et al, 1998; Danis et al, 2000; Metcalf et al, 2003, 2007). Thus, in the Queensland lungfish (Australian lungfish) *Neoceratodus forsteri*the albumin with the high level of the identity of NH_2- terminal fragment of 101 amino acids to the same fragment of mammalian albumin is discovered. The comparison of cloned albuminous cDNA of coelacanth (*Latimeriidae*) and Australian lungfish with the mammalian albumin genes confirmed the close relationship of these genes (Metcalf et al, 2003, 2007). The majority of the amino-acid sequence motives of NH_2-terminal albumin fragment from the white sturgeon albumin are identical to mammalian albumin (cited. Chikhachev, 1982).

The *Acipenseriformes* originated from the ancient representatives of *Paleonisci*, which relate to the actinopterygian fishes (*Actinopterygii*), but the sarcopterygiian fish (*Dipnoi*) and fleshy-finned fishes originated from ancient *Rhipidistia*, which gave beginning to the first tetrapods (*Tetrapoda*).

Meanwhile the albumins of *Acipenseriformes* are not similar to albumins of bony fishes, they are similar to albumins of the mammalian and contemporary representatives of sarcopterygiian fish (*Dipnoi*). This is the example of the convergent evolution of albumins in the posterities of *Rhipidistia* and *Paleonisci*.

```
P02769    MKWVTFISLLLLFSSAYSRGVFRRDTHKSEIAHRFKDLGEEHFKGLVLIAFSQYLQQCPF    60    ALBU_BOVIN
P02768    MKWVTFISLLFLFSSAYSRGVFRRDAHKSEVAHRFKDLGEENFKALVLIAFAQYLQQCPF    60    ALBU_HUMAN
P49065    MKWVTFISLLFLFSSAYSRGVFRREAHKSETAHRFNDVGEEHFPIGLVLITFSQYLQKCPY    60    ALBU_RABIT
P21848    MQKLSVCSLLVLLS------VLSRSQAQNQI:IIFTEAKEDGFKSLILVGLAQNLPDSTL    54    ALBUI_SALSA
          *:*::.  ***.*:*         *: *   .::: *.: **.*  ::** *   ...

P02769    DEHVKLVNELTEFAKTCVADESHAGCEKSLHTLFGDELCKVASLRFTYGDMADCCEKQEP    120   ALBU_BOVIN
P02768    EDHVKLVNEVTEFAKTCVADESAENCDKSLHTLFGDKLCTVATLRETYGEMADCCAKQEP    120   ALBU_HUMAN
P49065    EEHAKLVKEVTDLAKACVADESAANCDKSLHDIFGDKICALFSLRDTYGDVADCCEKKEP    120   ALBU_RABIT
P21848    GDLVPLIAEALAMGVKCCSDTPFEDCERDVADLFOSAVCSSETLVEKN-DLKMCCEKTAA    113   ALBUI_SALSA
          : *:* *       :*   .**:::  :* ..:* *  :  **  ..

P02769    ERNECFLSHKDDSP-DLPKLK-PDPNTLCDEFKADEKKFWGKYLYEIARRHPIFYAPELL    178   ALBU_BOVIN
P02768    ERNECFLQHKDDNI-NLFRLVRPEVDVMCTAPHDVEETFLKKYLYEIARRHPTFYAPELL    179   ALBU_HUMAN
P49065    ERNECFLHHKDDKP-DLPPFARPEADVLCKAFHDDEKAFFGHYLYEVARRHPTFYAPELL    179   ALBU_RABIT
P21848    ERTHCFVDHKAKIPRDLSLKAELPAADQCEDPKKDHKAFVGRFIPKFSKSNPMLPPHVVL    173   ALBUI_SALSA
          **.**  **  .:*,           * *: :*  *:::::**   *:..:.   :  :

P02769    YYANKYNGVFQECCQAEDKGACLLPKIETMREKVLASSARQRLRCASIQKFGERALKAWS    238   ALBU_BOVIN
P02768    FPAKRYKAAFTECCQAADKAACLLPKLDELRDEGKASSAKQRLKCASLQKFGERAFKAWA    239   ALBU_HUMAN
P49065    YYAQFYKAILTECCEAADKGACLTPKLDALESKSIISAAQERLRCASIQKFGDRAYKAWA    239   ALBU_RABIT
P21848    AIAKGYGEVLTTCCGEAEAQTCFDTKKATFQHAVMKRVAELRSLCIVHHKKYGDRVVKAKK    233   ALBUI_SALSA
          *:  *   :  **  .*    .*    :*     .   : .   :*:*:*.  **

P02769    VARLSQKFPKAEFVEVTKLVTDLTKVHKECCHGDLLECADDRADLAKYICDNQDTIS--S    296   ALBU_BOVIN
P02768    VARLSQRFPKAEFAEVSKLVTDLTKVHTECCHGDLLECADDRADLAKYICENQDSIS--S    297   ALBU_HUMAN
P49065    LVRLSQRFPKADFTDISKVITDLTKVHKECCHGDLLECADDRADLAKYNCEHQETIS--S    297   ALBU_RABIT
P21848    LVQYSQKMPQASFQEMGGMVDKIVATVAPCCSGDMVTCMKERKTLVDEVCADESVLSRAA    293   ALBUI_SALSA
          : : **::*:*.  ::   :*. . .    ** **::  .:*  *,.,.:*  .:.  :

P02769    KLKECCDKPLLEKSHCIAEVEKDAIPENLPPLTADFAEDKDVCKNYQEAKDAFLGSFLYE    356   ALBU_BOVIN
P02768    KLKECCEKPLLEKSHCIAEVENDEMPADLPSLAADFVESKDVCKNYAEAKDVFLGMFLYE    357   ALBU_HUMAN
P49065    HLKECCDKPILEKAHCIYGLNNDETPAGLFAVAEEFVEDKDVCKNYEEAKDLFLGKFLYE    357   ALBU_RABIT
P21848    GLSACCKEDAVHRCSCVEAMKFDFKFDCLSEHYDIHADIAAVCQTFTKTFDVAMCKLVYE    353   ALBUI_SALSA
          *.**.:   :.:.  *:     *   *  * . *.     ..:   **:: :.:  :* ::**

P02769    YSRRHPEYAVSVLLRLAKEYEATLEECCAKDDPHACYSFVPI--KLKHLVDEFQNLIKQN    414   ALBU_BOVIN
P02768    YARRHPDYSVVLLLRLAKTYETTLEKCCAAADPHECYAKVFI--EFKPLVEEFQNLIKQH    415   ALBU_HUMAN
P49065    YSRRHPDYSVVLLRLGKAYEATLKKCCATDDPHACYAKVLD--EFQPLVDEFKNLVKQN    415   ALBU_RABIT
P21848    ISVRHPESSQQVILRFAKEAEQALLQCCDMEDHAECVKTALAGSDIDKKITDETDYYKKM    413   ALBUI_SALSA
          :  :***:   :  :::**:.*  :*,**      **     :*    .:. .:  *:

P02769    CDQFEKLGEYGFQNALIVRYTRKVPQVSTPTLVEVSRSLGKVGTRCCTKPESE-RMPCTE    473   ALBU_BOVIN
P02768    CELFEQLGEYKFQNALLVRYTKKVPQVSTPTLVEVSRNLGKVGSKCCKHPEAK-RMPCAE    474   ALBU_HUMAN
P49065    CELYEQLGDYNFQNALLVRYTKKVPQVSTFTLVEISRSLGKVGSKCCKHPEAE-RLPCVE    474   ALBU_RABIT
P21848    CAAEAAVSDDSPEKSMKVYYTRINPQASPDQLHMVSETVHDVLHACCKEQGHFVLPCAE    473   ALBUI_SALSA
          *      .:    *:.:*:: ***:.*:  .*  .:**  .:.*   ..:*,  :*.*

P02769    DYLSLILNRLCVLHEKTPVSEKVTKCCTESLVNRKPCFSALTPDETYVPKAFDEKLFTPH    533   ALBU_BOVIN
P02768    DYLSVVLNQLCVLHEKTPVSDRVTKCCTESLVNRRPCFSALEVDETYVPKEFNAETFTFH    534   ALBU_HUMAN
P49065    DYLSVVLNRLCVLHEKTPVSEKVTKCCSESLVDRKPCFSALGFDETYVPKEFNAETFTFB    534   ALBU_RABIT
P21848    EKLTDAIDATCDDYDPSSINFHIAHCCNQSYSMRRHCILAIQFDTEFTFPELDASSPHMG    533   ALBUI_SALSA
          :*:  :: :    *  :: :.:.  :::::**.:*    **     *  :. :  ::

P02769    ADICTLPDTEKQIKKQTALVELLKHKPKATEEQLKTVMENFVAFVDKCCAADDKEACFAV    593   ALBU_BOVIN
P02768    ADICTLSEKERQIKKQTALVELVKHKPKATKEQLKAVMDDFAAFVEKCCKADDKETCFAE    594   ALBU_HUMAN
P49065    ADICTLPETERKIKKQTALVELVKHKPHATNDQLKTVVGEPTALLDKCCSAEDKEACFAV    594   ALBU_RABIT
P21848    EELCTKDSKDLLLSGKKLLYGVVVRHKTTITEDMLKTISTKYMTMKEKCCAAEDQRAGFTE    593   ALBUI_SALSA
          ..:**       :*     *: *****::    *::::: .:  ::  *** .*:*: .::*.

P02769    EGPKLVVSTQTALA-    607    ALBU_BOVIN
P02768    EGKKLVAASQAALGL    609    ALBU_HUMAN
P49065    EGPKLVESSKATLG-    608    ALBU_RABIT
P21848    EAPKLVSESAELVKY    608    ALBUI_SALSA
          *.  ***  :   :
```

Figure 8.1. The comparison of amino acid sequences of the albumins from Atlantic salmon*Salmo salar* (ALBU1_SALSA), human*Homo sapiens* (ALBU1_HUMAN), bull (ALBU1_BOVIN) and rabit (ALBU1_RABIT) (DB SwissProt, BLAST). The places of the proteolysis of albumin by tripsin marked with red.

The Origin of Albumin

Most probably, the albumins appeared in different taxa of fishes independently on the basis of the hemoglobin or myoglobin genes. The fragment of hemoglobin gene underwent the series of the alloyed duplications during epy evolution, which led subsequently to the appearance of the albuminous gene (White and other, 1978). Hemoglobins also appeared independently in the different groups of organisms on the basis of the cytochrome genes (Schmidt -Nielsen, 1079). If these hypotheses are valid, then homology of all albumins and albumin-similar proteins of all vertebrates can be considered as their indirect origin from the cytochromes.

8.2. PRINCIPLES OF THE FUNCTIONAL ORGANIZATION OF THE FISH BLOOD PLASMA PROTEINS

In many fresh-water bony fishes the erythrocytes have the increased tendency to the intravascular hemolysis, and after the penetration into the blood hemoglobin is extremely easily decomposed into the heme, the globin and Fe^{3+}. Under the unfavorable combination of the environmental factors the apoptosis of erythrocyte DNA can occur. It is accompanied by the decomposition of hemoglobin and by the ejection of the products of hemoglobin degradation into the blood flow. The operative system of the binding of the hemoglobin degradation products not only by the specialized proteins (transferrin, hemopexin and haptoglobin), but also by all unspecialized proteins of the blood helps the bony fishes to adapt and to survive under such conditions.

The structural stability of hemoglobins in cartilaginous and nonmigratory fresh-water *Acipenseriformes* fishes is low, but their erythrocytes are rather stable. Therefore there are no free hemoglobin and products of its destruction in the blood flow of such fishes, and it means and there is no need for the appearance of additional elements of the system of the binding of the ligands containing iron. In migratory *Acipenseriformes* fishes both erythrocytes and hemoglobin are stable. Hemoglobin maintains two-fold freezing, forming correct crystals after this procedure. Therefore in cartilaginous and *Acipenseriformes* fishes only separate elements of this system are discovered: in the form of specialized proteins transferrin and hemopexin (in cartilaginous fishes) or transferrin (in *Acipenseriformes* fishes); haptoglobins in the blood of these fishes are not found.

Thus, during the evolution of *Pisces* namely in *Teleostei* an increase in the number of specialized proteins, which bind intravascular hemoglobin and products of its destruction; and a decrease in the level of the specialization of all blood plasma proteins, which began to bind unspecifically all ligands containing iron, occurred. To the highest degree this tendency appeared in the serum oligomeric albumin, which binds hemin, and hemoglobin, and Fe^{3+}; i.e. it is a polyfunctional protein.

Thus, two polar types of the functional organization of the protein systems of the blood plasma in the fishes are revealed: 1) the differentiated systems from the specialized proteins in cartilaginous ganoids and 2) the differentiated systems from the weakly-specialized proteins in fresh-water bony fishes. The special features of the resistance properties of erythrocytes and the structural instability of hemoglobin are the factor, which determined the decrease in the level of the specialization of the blood plasma proteins. Reduction in the specialization of the proteins of fresh-water *Teleostei* is caused by physiological strategy of fish organism for averting of the losses of iron.

8.3. ROLE OF THE STRUCTURAL-FUNCTIONAL ORGANIZATION OF THE BLOOD PLASMA PROTEINS IN THE STABILIZATION OF THE INTERNAL FLUID ENVIRONMENT OF ORGANISM

The functions of the blood plasma proteins are accomplished at the direct participation of the vessels of circulatory system; the special features of the walls of vessels and the structural organization of the fish plasma proteins are adapted to each other:

1. In contrast to the mammals, in which the capillaries by the permeability to the proteins are differentiated from 0 to high values, in fishes the different type of capillaries are absolutely permeated to all blood plasma proteins. The selective nature of the permeability of the fish capillaries in relation to the different blood plasma proteins testifies to the participation of active mechanisms in the trans-capillary exchange of the proteins. The functional expediency of this selectivity is in the adaptation of metabolism in the different tissues of fishes to the conditions of environment and the physiological state of organism.

2. The stabilization of the internal fluid environment of organism is achieved not only by the regulation of the permeability of the wall of capillary, but also by the structural transformations of proteins, namely - oligomeric albumin, capable to the dissociation to the proteins (which enter the oligomer composition) during the trans-capillary exchange. Oligomeric albumins are discovered in the fresh-water representatives of the families *Esocidae, Cyprinidae* and *Percidae*, and also among the saltish-water fishes (*Clupeidae*). Among the marine fishes oligomeric albumin occurs rarely (*Scorpaena porcus*), however, its participation in the osmoregulation is not revealed (Andreeva, 2011; Andreeva et al., in of publ.).

The mechanism of the regulation of osmotic pressure with the help of the reversible dissociation of oligomeric proteins is characteristic for the intracellular proteins; for the blood plasma this mechanism is untypical (Shulz, Schirmer, 1979). However, this mechanism wasclaimed in the group of the fresh-water bony fishes, in which it works successfully in the conditions of the increased salinity of water.

3. The property of all blood plasma proteins to the specific and unspecific binding of the ligands containing iron, contributes to the stabilization of the internal fluid environment of organism. On the whole, it protects the internal fluids of organism from "the blockage" by the products of the hemoglobin decomposition.

CONCLUSION

The origin, formation and development of a whole series of proteins and protein families in the evolution of vertebral animals occur in the lowest vertebrates - fishes. The completion of the basic evolutionary transformations of these proteins in vertebral animals also occurs in *Pisces*, and no essential transformations of proteins in the subsequent classes of Vertebrata are observed. Since at the initial stages of the protein evolution their significant variety both in the structural and in the functional parameters is observed, the analysis of the structural and functional features of proteins in different taxa of *Pisces* makes it possible to approach understanding of basic appropriateness of their evolution in Vertebrata as a whole. The protein system of the blood plasma is of interest for such studies, as it is the specific acquisition and characteristic feature of all chord animals and participates in the adaptation of metabolism to the conditions of environment and the physiological state of the organism.

As stated above all features of similarity and differences in the blood plasma proteins of the *Pisces* and highest *Vertebrata* should be considered within the framework of their community. The features of the similarity of organization and integration of the blood proteins in the fishes and mammals reflect the relationship of all vertebrates, and differences are conditioned by their taxonomy and environmental conditions. Both in cartilaginous and in bony fishes the blood plasma proteins contain two basic fractions - albumins and globulins. However, only in bony fishes the protein composition of the blood is similar to that in higher vertebrates: among serum globulins of *Teleostei* not only immunoglobulins, transferrins and hemopexinare present, but also specialized protein haptoglobin, which is absent in cartilaginous and *Acipenseriformes* fishes.

The organization of immunoglobulin genes in the evolution of *Vertebrata* - from fishes to the mammals - underwent the transition from the cluster to the segmental model. In more ancient in evolutionary sense cartilaginous fishes the genes of immunoglobulins are organized in the form of repetitive clusters of V-J-C genic segments. The transition from the cluster organization of IG genes to the segmental one occurred in *Osteichthyes* (*Acipenseriformes* and bony fishes): there are IGL genes in *Acipenseriformes* and IGH ones in bony fishes are organized according to the segmental type. In mammals all immunoglobulin system is organized according to the segmental type, which ensures the maximal variety of immune response due to the combinatorial recombination of gene segments. As for other globulins - transferrin, hemopexin and haptoglobin - the information about their significant difference in the way of organization of these proteins in the fishes (according to monomeric type) and in mammals is absent. The conservatism in the organization of these proteins is explained, probably, by the importance of their specific functions of the binding of the iron-containing ligands for the purpose of preventing their losses by organism.

One of the most variable fractions of serum proteins in fishes is the low-molecular fraction, which includes albumins and other low-molecular proteins. In different taxa and biotopes these fractions have their unique protein composition and properties. Only in the *Dipnoi* and *Acipenseriformes* fishes the albumins contain motives in their structure similar to the same in mammal albumins. The other fishes have albumins and albumin-similar proteins, which primary and surface structures differ not only from mammal albumins, bur from other fish ones. In mammals the low-molecular protein fraction is represented by large specialized proteins, which consist of one polypeptide chain (monomers), which are filtered through the wall of capillary in the specific divisions of the capillary bed. In fishes among the proteins of low-molecular protein fraction the simple and complex proteins, monomers and oligomers, the highly-heterogeneous and not- heterogeneous system are detected. All their elements are filtered in all divisions of capillary network. However, the principle of the unity of this group of proteins in all vertebrates is observed and is manifested under the conditions of the pathology of mammals, when their serum proteins become similar to the proteins of fishes.Thus, human serum albumin at the disease by diabetes and some other pathologies (bisalbuminemia) greatly resembles the albumins of the fresh-water bony fishes: such albumin is glycosilated, aggregated and behaves anomalously in the electrophoresis, on the electrophoregram it can be in the form of a spot with indistinct outlines, etc. Under conditions of pathology an increase in the permeability of capillaries to the proteins is observed in mammals and such permeability is the norm for fishes. The similarity in the organization of the proteins of mammals (under conditions of pathology) and fishes removes distinct boundaries between them. It makes it possible to trace indirectly the stages of the formation of the blood protein organization in the *Vertebrata* evolution.

The organization of the blood proteins and their distribution in extracellular fluid compartments were formed in fishes in accordance with the special features of their internal environment and habitat. The salinity is one of the basic limiting factors for aquatic animals. Depending on selected strategy by fishes - to the hyper- or hypoosmoticinternal fluids of organism relative to sea water - extracellular fluids of organism with the unique composition, that preserved one level of salinity, somewhat higher than 5-8‰ - were formed (Khlebovich, 1974). These extracellular fluids determined the configuration of the protein systems of the blood plasma characteristic for each taxon and biotope, which is observed in cartilaginous, *Acipenseriformes*, marine and fresh-water bony fishes. In cartilaginous fishes highly differentiated protein systems of the blood plasma were formed; the high content of urea in the fluids of organism supports the unaggregated form (monomers) of their proteins. In cartilaginous ganoids the highly differentiated blood protein systems from the specialized proteins were formed, their low-molecular fractions were represented exceptionally by monomer albumins. The monomeric proteins in the composition of low-molecular fractions are typical for marine bony fishes as well. However, in this group of fishes the species are revealed, in low-molecular fractions of which the oligomeric proteins are detected. The protein systems of the blood plasma of the fresh-water bony fishes are organized the most complicatedly: they are formed from the polyfunctional monomer and oligomeric proteins, which consist of 10 or more polypeptide chains; the proteins are capable to penetrate through the wall of capillary by means of the mechanism of selective permeability into the tissue fluids, in which the oligomers can dissociate to the polypeptide chains during the adaptations to the indices of salinity boundary for the fresh-water fishes. The oligomerization of the blood plasma proteins observed in the group of fresh-water *Teleostei* is accompanied by the decrease in the level of their specialization. The special features of the molecular organization of hemoglobin are the factor, which determines such decrease.

The appearance of oligomeric proteins in the blood of fresh-water *Teleostei* can explained by the functional expediency of this acquisition. The dehydration does not threaten the proteins of fresh-water fishes by reason of hypertonicity of their internal biological fluids, vice - versa – the organism constantly pumps out the excess water, and the formation of the protein complexes can decrease the oncotic pressure of these fluids, which contributes to the rapid redistribution of extracellular water and, as a whole, to stabilization of water exchange. It is advantageous for sea species under conditions of hypertonic environment to have many small monomeric proteins in blood in order to retain water in the organism and thus to avoid the thread of dehydratation, because the formation of protein aggregates decreases the oncotic pressure of the blood and aggravates the problem of dehydration.

The evolutionary past of Pisces is supposedly connected with the prolonged initial stage of life in the sea water(Romer, Parsons, 1992). The subsequent stage of the adaptation to fresh waters by marine species required the serious reconstructions of the water exchange of fishes, including the processes of filtration. Taking into account the formation of primary protein systems with the salinity of the internal environment of approximately 5-8‰, it can be assumed that the proteins of ancient marine fishes could exist in the internal fluid environment in the form of the separate polypeptide chains, which during the fish adaptation to fresh waters were united into the complexes, which contributed to reduction in the value of the oncotic pressure of the blood and stabilized, as a whole, the processes of filtration in the organism of fishes under the conditions of fresh waters. The special feature of the surface structure of the proteins in the form of carbohydrate residues enabled the integration of fish proteins into the complexes. Subsequently such protein complexes lost their initial value in the fishes, which moved from the sea water into the fresh waters and became to live according to the nonmigratory mode of life. However, the mechanism, reverse to association, namely the mechanism of the dissociation of serum oligomeric albumin, proved to be useful for the regulation of the processes of filtration inside the organism, and also convenient during the adaptations of plastic exchange (the ripening of the gonads, under the conditions of the shortage of food). In the both cases - during the adaptations and water and plastic metabolism - the universal algorithm of the structural transformations of oligomeric albumin according to the type of its dissociation for subunits and redistributions of the latter between the intra- and extravascular spaces is used.The presence of oligomeric albumins in the blood of fresh-water bony fishes can be considered as the ancient feature, which resembles about the stage of the fish migration from the marine into the fresh-water environment.

The acquisition by the fresh-water bony fishes of the mechanism of the rapid "trimming" of the oncotic pressure of extracellular fluids to optimum values *in situ* due to the dissociation of oligomeric complexes doesn't require any serious reconstructions of fish genome (in contrast to immunoglobulins, which changes were determined by transition of the cluster organization of IG genes in cartilaginous fishes to the segmental one in *Osteichthyes*) or of physiological functions; onlyspecial features of the surface structure of the proteins, which contain carbohydrate residues, and the lability of the proteins structure could contribute to them. Such protein labilitypresents the great possibilities for their conformational conversions under the action of different factors, which lead to association or dissociation of protein molecules, and can be used in the osmoregulation.

In mammals the mechanism of osmoregulation with the participation of the oligomeric proteins is characteristic only for the intracellular fluids. In fresh-water bony fishes this mechanism proved to be necessary for the extracellular fluids.The realization of this mechanism in fishes is provided and facilitated by the high permeability of their capillaries to all serum proteins. Due to this mechanism the blood plasma protein systems of fresh-water bony fishes acquired dynamic nature and the bony fishes maximally optimized their water-salt and plastic exchange. This could not but increase the chances of this group of fishes to the exploit new biotopes and as a whole determine their ecological and evolutionary success.

REFERENCES

Adamson, RH; Lenz, JF; Zhang, X; Adamson, GN; Weinbaum, S; Curry, FE. Oncotic pressures opposing filtration across non-fenestrated rat microvessels. *J.Physiol.*, 2004 557(3), 889-907.

Aisen, P; Leibman, A; Zweier, J. Stoichiometric and site characteristics of the binding of iron to human transferring. *Birth.Defects.Orig.Artic.Ser.*, 1978 12(8), 81-95.

Alexeev, GA; Berliner, GB. *Hemoglobinuria*. Moscow: Medicine Publisher; 1972.

Alexandrov, VY. *Reactivity of cells and the proteins*. Leningrad: Nauka Publisher; 1985.

Ali Han, MV; Rashid, Z; Ali Han, V; Ali, R. Biochemical, biophysical and thermodynamic analysis of human serum albumin glycated in vivo. Original Russian text is published in *Biochemistry*, 2007 72(2), 175-183.

Altruda, F; Poli, V; Restagno, G; Argos, P; Cortese, R; Silengo, L. The primary structure of human hemopexin deduced from cDNA sequence: evidence for internal, repeating homology. *Nucleic Acid Res.*, 1985 13(11), 3841-59.

Anderson, MK; Shamblott, MJ; Litman, RT; Litman, GW. Generation of immunoglobulin light chain gene diversity in Raja erinacea is not associated with somatic rearrangement, an exeption to a central paradigm of B cell immunity. *Joutnal of Experimental Medicine*, 1995 182, 109-119.

Anderson, WG; Takei, Y; Hazon, N. Osmotic and volaemic effects on drinking rate in elasmobranch fish. *J.Exp.Biol.*, 2002 205(8), 1115-22.

Anderson, WG; Taylor, JR; Good, JP; Hazon, N; Grosell, M. Body fluid volume regulation in elasmobranch fish. *Comp. Biochem. Physiol. A.Mol. Integr. Physiol.*, 2007 148(1), 3-13.

Andreeva, AM. Similarity and difference in the binding of bromcresol purple by serum albumin of the sturgeon and bony fishes. In: *Abstract book (VI Ecological fish biochemistry Conf.).* Vilnus; 1985; 6.

Andreeva, AM. Identification of serum albumin and the study of some of its physical and chemistry properties in the representatives of the families Acipenseridae and Cyprinidae. *Inf.Bull.IBII AS USSR*, 1986a 69, 36-39.

Andreeva, AM. Physical and chemical properties of serum proteins from cartilaginous fishes on the example spiny dogfish. In: *Matherials of IX All-Russian conference of evolutional physiology.* Leningrad: Nauka; 1986b; 13-14.

Andreeva, AM. Identification of serum transferrins of the bream and starlet. In: Lukjanenko VI. *Matherials of I Symp. of Ecological biochemistry of the fishes.* Yaroslavl: YrGU; 1987a; 8-10.

Andreeva, AM. About the role of sialic acids in the creation of the heterogeneity of transferrins of sterlet and bream. In: Lukjanenko VI. *Matherials of I Symp. of Ecological biochemistry of the fishes.* Yaroslavl: YrGU; 1987b; 10-11.

Andreeva, AM. Physical and chemical properties of serum albumin of the blood from Acipenseridae and Cyprinidae on the example to sterled and bream. In: Saburov EG. *Physiology and the biochemistry of the hydrobionts.* Yaroslavl: YrGU; 1987c; 108-114.

Andreeva, AM. The stability of hemoglobin ofAcipenseridae to the dehydrating action of ammonium sulfate. *Inf.Bull.IBII AS USSR*, 1987d 76, 56-59.

Andreeva, AM. About the structure of hemoglobin of some species of the family Acipenseridae. *Inf.Bull.IBII AS USSR*, 1987e 75, 33-36.

Andreeva, AM. Physical and chemical properties of the basic groups of the blood proteins in the different ecological and taxonomic groups of cartilaginous, cartilaginous ganoids and bony fishes: dissertation abstract...candidate of biological science. Borok: IBII RAS; 1997; 1-24.

Andreeva, AM. Structural and functional organization of albumin system of fish blood. *Journal of Ichthyology*, 1999 39(9), 788-794. Original Russian text is published in *Voprosy Ikhtiology*, 1999 39(6), 825-832.

Andreeva, AM. Serum peroxidases of fish. *Journal of Ichthyology*, 2001a 41(1), 104-111. Original Russian text is published in *Voprosy Ikhtiology*, 2001 41(1), 113-121.

Andreeva, AM. Serum gamma-globulins of the fishes. *Journal of Ichthyology*, 2001b 41(6), 464-470. Original Russian text is published in *Voprosy Ikhtiology*, 2001 41(4), 550-556.

Andreeva, AM. Formation of Isozyme Spectra of Lactate Dehydrogenase During Early Development of Roach *Rutilus rutilus* (L.). *Journal of Ichthyology*, 2005a 45(2), 205-209. Original Russian text is published in *Voprosy Ikhtiologii*, 2005 45(2), 277-282.

Andreeva, AM. Specific features of expression of lactate dehydrogenase genes in early development of the bream Abramis brama (L.), roach Rutilus rutilus (L.) and their reciprocal F1 hybrids. *Journal of Ichthyology*, 2005b 45(4), 334-339. Original Russian text is published in Voprosy Ikhtiologii, 2005 45(3), 411-417.

Andreeva, AM. Influence of the destabilizing factors on the structural-functional indices of hemoglobin of the nonmigratory and migratory fishes. Original Russian text is published in *Zhurnal Evolyutsionnoi biokhimii I fiziologii*, 2006 42(6), 537-543.

Andreeva, AM; Yurkova, MS; Rjabtseva, IP; Lukjanenko, VV; Sharapova, OA; Kuzmina, VA. Maintenance of the organism homeostasis of the goldfish *Carassius auratus gibellio* (Bloch) under the conditions of the unfavorable combination of some factors. In: Ivanov DV. *Actual problems of hydroecology*. Kazan: Otechestvo; 2006; 215-220.

Andreeva, AM. Specific features of expression of aspartat aminotransferase genes in early development of some cyprinid fishes and their intergeneric F1 hybrids. *Russian Journal of Developmental Biology*, 2007 38(1), 35-41. Original Russian text is published in *Onthogenez*, 2007 38(1), 44-51.

Andreeva, AM; Chalov, JP; Ryabtseva, IP. Peculiarities of distribution of plasma proteins between the internal medium specialized compartments in the carpCyprinus carpio L. Original Russian text is published in *Journal of Evolutionary biochemistry and physiology*, 2007 43(6), 825-832.

Andreeva, AM; Ryabtseva, IP; Bolshakov, VV. Analysis of permeability of capillaries of various parts of the microcirculatory system to plasma proteins in some representatives of bony fish. *Journal of Evolutionary biochemistry and physiology*, 2008 44(2), 251-253. Original Russian text is published in *Zhurnal Evolyutsionnoi biokhimii I fiziologii*, 2008 44(2), 212-213.

Andreeva, AM. Structural and functional organization of blood and some extracellular fluids of the fishes: dissertation abstract...doctor of biological science. Moscow: Moscow State University (MSU); 2008; 1-45.

Andreeva, AM. The role of internal fluid environment for regulation of germ genes expression in early development of some cyprinid fishes and their intergeneric F1 hybrids. In: Kurzfield NC. *Developmental Gene Expression Regulation*.N.Y.: Nova Science Publisher Inc.;2009; 263-283.

Andreeva, AM; Ryabtseva, IP; Lukjanenko VV. Adaptations of the respiratory function of the blood in the fresh-water bony fishes. *Matherials of 28th Int.conf. "Biological resources of the White Sea and internal reservoirs of European North"*. Petrozavodsk: Karelian SC RAS; 2009; 33-39.

Andreeva, AM. The structure of serum albumins of fishes. Original Russian text is published in *Zhurnal Evolyutsionnoi biokhimii I fiziologii*, 2010a 46(2), 111-118.

Andreeva, AM. The role of structural organization of blood plasma proteins in the stabilization of water metabolism in bony fish (Teleostei). *Journal of Ichthyology*, 2010b 50(7), 552-558. Translated from *Voprosy Ikhtiology*, 2010 50(4), 570-576.

Andreeva, AM; Fedorov, RA. Features of the organization of low-molecular weight proteins from the blood and tissue fluid of the common stingray Dasyatis pastinacaL. (Chondroichthyes: Trigonidae). Russian Journal of marine biology, 2010 36(6), 469-472. Translated from Biologiya moray, 2010 36(6), 460-462.

Andreeva, AM; Fedorov, RA; Ryabtseva, IP. Comparative analysis of filtration mechanismes in freshwater and seawater bony fishes. *Matherials of VI scientific conference"Actual problems of ecology"*. Grodno: state-run university in Grodno; 2010; 163-165.

Andreeva, AM; Dmitrieva, AE. The diversity of heavy chain of immunoglobulins of scorpionfish Scorpaena porcus L. *Matherials of III scientific conference of the problem of immunology, pathology and protection of the health of the fishes and other hydrobionts*. Borok: IBIW RAS; 2011;81-85.

Andreeva, AM; Ryabtseva, IP. Mechanisms of respiratory function of the blood in Teleostei. Original Russian text is published in *Voprosy Ikhtiology,* 2011 51(5).

Andreeva, AM. Participation of Yolk Proteins in Regulation of Germ Genes Expression from Intergeneric F1 Hybrids of Bream, Roach and Blue Bream. *Int. Journ. Med. Biol. Fr.*, 2011 17(1/2).

Andreeva, AM. Mechanisms of the plurality of *Scorpaena porcus* L. serum albumin. *Open Journal of Marine Science*. 20111; 31-35. doi:104236/ojms.2011.12003 Published Online July 2011 (http://www.SciRP.org/journal/ojms).

Andreeva, AM, Ryabtseva, IP; Fedorov, RA. Organization of low molecular proteins of blood and tissue fluid from marine bony fishes. 2011. In of publ.

Arai, K; Huss, K; Madison, J; Putnam, FW. Point substitutions in albumin genetic variants from Asia. *Proc. Natl. Acad. Sci. U.S.A.* 1990 87; 497-501.

Armour, KJ; O'Toole, LB; Hazon, N. The effect of diet protein restriction on the secretory dynamics of la-hydroxycorticosterone and urea in the dogfish, Scyliorhinus canicula: a possible role for la-hydroxycorticosterone in sodium retention. *J.Endocrinol.*, 1993 138, 275-282.

Balahnin, IA; Galagan, NP; Lukjanenko, VI; Popov, AV. Genetic polymorphism on some components of the blood of fishes (sturgeon and carp). *DASUSSR*, 1972 204(5), 1250-1252.

Bartl, S; Weissman, IL. PCR primers containing an inosine triplet to complement a variable codon within a conserved protein-coding region. *J.BioTechniques*, 1994 16(2), 246-248.

Bajunova, LV; Barannikova, IA; Dubin, VP; Semenkova, TB. Hormonal characteristics of sturgeon under the conditions of the stress. *Matherials of Intern.Confer. "Sturgeon fishes at the turn of the 21th century*. Astrakhan; 2000; 122-123.

Bede, M. The proteins and lipids of plasma of some species of Australian fish and water salt fish. *J.Cell.Physiol.*, 1959 54(3), 221-230.

Behrens, PG; Spikerman, AM; Brown, JB. Structure of human serum albumin. *Fed.Proc.*, 1974 34, 2106.

Bendayan, M; Rasio, EA. Transport of insulin and albumin by the microvascular endothelium of the rete mirabile. *J.Cell.Sci.*, 1996 109(7), 1857-64.

Bendayan, M; Rasio, EA. Evidence of a tubular system for transendothelial transport in arterial capillaries of the rete mirabile. *J.Histochem.Cytochem.*, 1997 45(10), 1365-78.

Bentley, PJ. A high-affinity zinc-binding plasma protein in channel catfish (Ictalurus punctatus). *Comp. Biochem. Physiol.*, 1991 100(3), 491-494.

Brand, S; Hutchinson, DW; Donaldson, D. Albumin Redhill, a human albumin variant. *Clin. Chim. Acta.* 1984 136, 197-202.

Brennan, SO; Myles, T; Peach, RJ; Donaldson, D; Georg, PM. Albumin Redhill (-1 Arg, 320 Ala–Thr): a glycoprotein variant of human serum albumin whose precursor has an aberrant signal peptidase cleavage site. *Proc. Natl. Acad. Sci. U.S.A.* 1990 87, 26-30.

Brown, GW; Brown, SG. Urea and its formation in coelacanth liver. *Science.* 1967 155, 57-63.

Brown, WM; Dziegielewska, KM; Foreman, RC; Saunders, NR. Nucleotide and deduced amino acid sequence of sheep serum albumin. *Nucleic Acids Res.*, 1989 17(24), 10495.

Bundgaard, M; Cserr, HF. Impermeability of hagfish cerebral capillaries to radio-labelled polyethylene glycols and to microperoxidase. *Brain Res.*, 1981 206(1), 71-81.

Bundgaard, M. Brain barrier systems in the lamprey. I. Ultastructure and permeability of cerebral blood vessels. *Brain Res.*, 1982 240(1), 55-64.

Burger, JW. Roles of the rectal gland and the kidneys in salt and water excretion in the spiny dogfish. *Physiol.Zool.*, 1965 38, 191-196.

Burmester, T; Ebner, B; Weich, B; Hankeln, T. Cytoglobin: a novel globin type ubiquitously expressed in vertebrate tissues. *Mol. Biol. Evol.*, 2002 19, 416–421.

Burmeste, T; Hankeln, T. Neuroglobin: A respiratory protein of the nervous system. *News Physiol. Sci.*, 2004 19, 110–113.

Byrnes, L; Gannon, F. Atlantic salmon (Salmo salar) serum albumin: cDNA sequence, evolution, and tissue expression.*DNA,* 1990 9(9), 647-565.

Campagnoli, M; Rosipal, S; Debreova, M; Rosipal, R; Sala, A; Romano, A; Labo, S; Galliano, M; Minchiotti, L. Analbuminemia in a Slovak Romany (gypsy) family: case report and mutational analysis. *Clin. Chim. Acta.* 2006 365, 188-193.

Carlson, J; Sakamoto, Y; Laurell, CB; Madison, J; Watkins, S; Putnam, FW. Alloalbunemia in Sweden: structural study and phenotypic distribution of nine albumin variants. *Proc. Natl. Acad. Sci. U.S.A.* 1992 89, 8225-8229.

Chalov, YP; Lukjanenko, VI. Special features of the fractional composition of the proteins of interstitial fluid and blood plasma in some *Cyprinidae* species. In:Lukjanenko VI. *Matherials of VII Confer. of Ecological physiology and biochemistry of the fishes.* Yaroslavl: YrGU; 1989; 214-215.

Chihachev, AS. Use of biochemical markers in the sturgeon economy of the Azov basin. Manuscript deposited in All-Russian Institute of Scientific and Technical Information, №3845-82 DEP. Rostov-on-Don university. 1982.

Chihachev, AS. Control of the genetic structure of populations and the hybridization of the valuable species of fishes during the artificial breeding. In: *Biological bases of fish-breeding; the problem of genetics and selection.* Leningrad: Nauka; 1983; 91-102.

Chihachev, AS; Tsvetnenko, YB. Study of the proteins of the blood of Azov sturgeon with their artificial reproduction. *Trudi Vsesojuznogo NII rybnogo hozjajstva I okeanografii*, 1979a 52(1), 87-182.

Chihachev, AS; Tsvetnenko, YB. Polymorphism of albumin and transferrin in the Azov population of the sturgeon (*Acipenser stellatus* L.). In: *Biochemical and population genetics of the fishes*. Leningrad: Nauka; 1979b; 111-115.

Chutaeva, AI; Dombrovskij, VK; Guzuk, SN; Lazovskij, AA. Polymorphism isobelinskiy's carp along some protein systems. In: *Bases of the bioproductivity of internal waters of the Baltic States*. Vilnus; 1975; 311-313.

Cordier, D; Barnoud, R; Branndonn, AM. Etude sur la proteinemie de la Roussette (Scyllium canicula L.). Influence du jeune. C. R. Soc.Biol.(Paris), 1957 151, 1912-1915.

Curry, FR. Microvascular solute and water transport. *Microcirculation*, 2005 12(1), 17-31.

Daggfeldt, A; Bengten, E; Pilstrom, L. A cluster type organization of the loci of the immunoglobulin light chain in Atlantic cod (Gadus morhua L.) and rainbow trout (Oncorhynchus mykiss Walbaum) indicated by nucleotide sequences of cDNAs and hybridization analysis.*Immunogenetics*, 1993 38(3), 199-209.

Danis, MH; Filosa, MF; Youson, JH. An albumin-like protein in the serum of non-parasitic brook lamprey (Lampetra appendix) is restricted to preadult phases of the life cycle in contrast to the parasitic species Petromyzon marinus. *Comp. Biochem. Physiol., B., Biochem. Mol. Biol.*, 2000 127(2), 251-260.

Denisov, SG. Cloning and the analysis of the genes of light chains of the immunoglobulins of the sterlet: diploma thesis. Novosibirsk: Novosibirsk state-run university, Institute of cytology and genetics of Siberia branch of Russian Academy of Sciences; 1997; 1-52.

Drilhon, A; Fine, JM. Les proteins du serum sanguine chez les Elasmobranches (Scyllium catulus et Scyllorhinus canicula). *C.R. Acad. Sci.*(Paris), 1959 248, 2418-2420.

Duff, DW; Olson, KR. Response of rainbow trout to constant-pressure and constant-volume hemorrhage. *Am. J. Physiol.*, 1989 257(2), 1307-14.

Dugaiczyk, A; Law, SW; Dennison, OE. Nucleotide sequence and the encoded amino acids of human serum albumin mRNA. *Proc. Natl. Acad. Sci. U.S.A.* 1982 79(1), 71-5.

Fainshtein, FE; Kozinec, GI; Bahramov, SM; Hohlova, MP. *Diseases of the system of the blood.* Tashkent: Medicine; 1987.Fedorov, PA; Andreeva, AM. Special features of the trans-capillary exchange of the proteins of the plasma of the blood in the fresh-water bony fishes. *Matherials of XXVIII Scientific conference "Biological resources of the White Sea and internal reservoirs of the European north".* Petrozavodsk: Karelian scientific center RAS; 2009; 583-585.

Fedorov, PA; Andreeva, AM. Special features of the trans-capillary exchange of the proteins of the plasma of the blood in the sterlet *ACIPENSER RUTHENUS* L. In: *Matherials of Scientific conference "Ecological studies in the region of Russia and in the contiguous territories".* Saransk; 2010; 186-188.

Fellows, FC; Hird, FJ; Fatty acid binding proteins in the serum of various animals. *Comp.Biochem.Physiol.B.*, 1981 68B, 83-87.

Fevolden, SE; Ried, KH; Fjalestad, KT. Selection response of cortisol and lysozyme in rainbow trout and correlation to growth. *Aquaculture*, 2002 205(1-2), 61-75.

Filosa, MF; Adam, I; Robson, P; Heinig, JA; Smith, K; Keeley, FW; Youson, JH. Partial clone of the gene for AS protein of the lamprey Petromyzon marinus, a member of the albumin supergene family whose expression is restricted to the larval and metamorphic phases of the life cycle. *J. Exp. Zool.*, 1998 282(3), 301-309.

Flajnik, MF. The immune system of ectothermic vertebrates. *Veterinary Immunology and immunopathology*, 1996 54, 145-150.

Flouriot, G; Ducouret, B; Byrnes, L; Valotaire, Y. Transcriptional regulation of expression of the rainbow trout albumin gene by estrogen. *J. Mol. Endocrinol.*, 1998 20(30), 355-362.

Freitas, TA; Hou, S; Dioum, EM; Saito, JA; Newhouse, J; Gonzalez, G; Gilles-Gonzalez, MA; Alam, M. Ancestral hemoglobins in Archaea. *Proc. Natl. Acad. Sci. USA.*, 2004 101, 6675–6680.

Fyhn, UEH; Bolling, S. Elasmobranch hemoglobins; dimerization and polymerization in various species. *Comp. Biochem. and Physiol. B.*, 1975 50(1), 119-129.

Gaal, O; Medgyesi, GA; Vereczkey, L. Electrophoresis in the Separation of Biological Macromolecules. Moscow: Mir Publisher. 1982. Translated from Inglish.

Galindez, EJ; Aggio, MC. Thespleen of Mustelus schimitti (Chondrichthyes, Triakidae): a light and electron microscopic study. *Ichthyol.Res.*, 1998 45, 179-186.

Galliano, M; Campagnoli, M; Rossi, A; Witsing von Konig, CH; Lyon, AW; Cefle, K; Yildiz, A; Palanduz, S; Ozturk, S; Minchiotti, L. Molecular diagnosis of analbuminemia: a novel mutation identified in two Amerindian and two Turkish families. *Clin. Chem.*, 2002 48, 844-849.

Gamble, JL. *Chemical Anatomy, Physiology and Extracellular Fluid.* 6th ed. Harvard University Press. Cambridge. Mass.; 1954.

Ghaffari, SH; Lobb, CJ. Structure and genomic organization of immunoglobulin light chain in the channel catfish. An unusual genomic organizational pattern of segmental genes. *J.Immunol.*, 1993 151(12), 6900-6912.

Ghaffari, SH; Lobb, CJ. Structure and genomic organization ofa second class of immunoglobulin light chain genes in the channel catfish. *J.Immunol.*, 1997 159(1), 250-258.

Goldstein, L; Forster, RP. Osmoregulation and urea metabolism in the little skate, Raja erinacea. *J.Physiol.*, 1971 220, 742-746.

Goldstein, LA; Heath, EC. Nucleotide sequence of rat haptoglobin cDNA. Characterization of the alpha beta- subunit junction region of prohaptoglobin. *J.Biol.Chem.*, 1984 259, 9212-9217.

Greenberg, AS; Steiner, L; Kasahara, M; Flajnik, MF. Isolation of a immunoglobulin light chain cDNA clone encoding a protein resembling mammalian k light chains: Implications for the evolution of light chains.*Proc.Natl.Acad.Sci.U.S.A.* 1993 90, 10603-10607.

Griffith, RW; Umminger, BZ; Grant, BF; Pang, PKT; Picford, JE. Serum composition of the coelacanth, Latimeria chalumnae Smith. *J.Exp.Zool.,* 1974 187(1), 87-102.

Grove, S; Tryland, M; Press, CM; Reital, LJ. Serum immunoglobulin M in Atlantic halibut (Hippoglossus hippoglossus): characterization of the molecule and its immunoreactivity. *Fish.Shellfish Immunol.*, 2006 20(1), 97-112.

Guyton, AC; Granger, HJ; Taylor, AE. Interstitial Fluid Pressure. *Physiological Reviews.* 1971 51(3), 527-563.

Haefliger, DN; Moskaitis, JE; Schoenberg, DR; Wahli, W. Amphibian albumins as members of the albumin, alpha-fetoprotein, vitamin D-binding protein multigene family. *J.Mol.Evol.*, 1989 29(4), 344-54.

Hansen, JD; Landis, ED; Phillips, RB. Discovery of a unique Ig heavy-chain isotype (IgT) in rainbow trout: Implication for a distinctive B cell developmental pathway in teleost fish. *PNAS*, 2005 102(19), 6919-6924.

Hargens, AR; Millard, RW; Johansen, K. High capillary permeability in fishes. *Comp. Biochem. Physiol. A Comp. Physiol.*, 1974 48(4), 675-80.

Harms, C; Ross, T; Segars, A. Plasma biochemistry reference values of wild bonnethead sharks, Sphyrna tiburo. *Vet. Clin. Pathol.,* 2002 31(3), 111-115.

Harris, DC; Aisen, P. *Iron carries and iron proteins.* VCH Publishers. New York; 1998.

Hawkins, JW; Dugaiczyk, A.The human serum albumin gene: structure of a unique locus. *Gene.* 1982 9(1), 55-8.

Hazon, N; Tierney, ML; Takei, Y. Renin – angiotensin system in elasmobranch fish: A review. *J.Exp.Zool.,* 1999 284, 526-534.

He, XM; Carter, DC. Atomic structure and chemistry of human serum albumin. *Nature.* 1992 358, 209-215.

Higgins, D; Thompson, J; Gibson, T; Thompson, JD; Higgins, DG; Gibson, TJ. CLUSTAL W: improving the sensitivity of progressive multiple sequence alignment through sequence weighting,position-specific gap penalties and weight matrix choice. *Nucleic Acids Res.* 1994 22, 4673-4680.

Hirayama, K; Akashi, S; Furuya, M; Fukuhara, K. Rapid confirmation and revision of the primary structure of bovine serum albumin by ESIMS and Frit-FAB LC/MS. *Biochem.Biophys.Res.Commun.* 1990 173, 639-646.

Ho, JX; Holowachuk, EW; Norton, EJ; Twigg, PD; Carter, DC. X-ray and primary structure of horse serum albumin (Equuus caballus) at 0,25-nm resolution. *European Journal of Biochemistry.* 1993 215(1), 205-212.

Hohman, VS; Schluter, SF; Marchalonis, JJ. Complete sequence of a cDNA clone specifying sandbar shark immunoglobulin light chain: Gene organization and implication for the evolution of light chains. *Proc.Natl.Acad.Sci.U.S.A.* 1992 89, 276-280.

Hohman, VS; Schuchman, DB; Schluter, SF; Marchalonis, JJ. Genomic clone for sandbar shark light chain: Generation diversity in the absence of rearrangement. *Proc.Natl.Acad.Sci.U.S.A.* 1993 90, 9882-9886.

Hsu, E; Pulham, N; Rumfelt, L; Flajnik, MF. The plasticity of immunoglobulin gene system in evolution. *Immunol.Rev.* 2006 210, 8-26.

Hyodo, S; Bell, JD; Healy, JM; Kaneko, T; Hasegawa, S; Takei, Y; Donald, JA; Toop, T. Osmoregulation in elephant fish Callorhinchus milii (Holocephali), wich special reference to the rectal gland. *J.Exp.Biol.* 2007 210 (8), 1303-1310.

Hutchinson, DW; Matejtschuk, P. The N-terminal sequence of albumin Redhill, a variant of human serum albumin. *FEBS Lett,* 1985 193, 211-212.

Idelson, LI; Didkovskij, ND; Ermilchenko, GV. *Hemolytic anemias.* Moscow: Medicine; 1975.

Jameson, GB; Anderson, BF; Norris, GE; Thomas, DH; Baker, EN. Structure of human apolactoferrin at 2.0A resolution. Refinement and analysis of ligand-induced conformational change. *Addendum Acta Crystallogr. D Biol. Crystallogr.* 1999 55, 1108.

Jeffrey, PD; Bewley, MC; MacGillivray, RT; Mason, AB; Woodworth, RC; Baker, EN. Ligand-induced conformational change in transferrins: crystal structure of the open form of the N-terminal half-molecule of human transferring. *Biochemistry*, 1998 37(40), 13978-86.

Jeong, JY; Kwon, HB; Ahn, JC; Kang, D; Kwon, SH; Park, JA; Kim, KW. Functional and developmental analysis of the blood-brain barrier in zebrafish. *Brain Res. Bull.*, 2008 75(5), 619-28.

Junko, K; Takaji, I. Cortisol directly inhibits neutrophil defense activities in tilapia. In: *Book Abstr. 9th Int. Conf. "Diseases Fish and Shellfish".* Rhodes; 1999; 293.

Karnauhov, GI. Haptoglobins in some fishes of the Far-Eastern complex. *Inf.Bull.IBII AS USSR*, 1987 75, 53-55.

Kejvanfar, A. Serologie et immunologie de deux especes de thonides (Germo alalunga Gm. Et Thunnus thynnus L.) de l Atlantique et de la Mediterranee. *Rev. Trav. Inst. Peches mar.*, 1962 26(4), 407-450.

Khlebovich, VV. The critical salinity of biological processes. Publ. *"Nauka"*, Leningrad. 1974.

Kiernan, JA; Contestabile, A. Vascular permeability accociated with axonal regeneration in the optic system of the goldfish. *Acta Neuropathol.*, 1980 51(1), 39-45.

Kiron, V; Takeuchi, T; Watanabe, T. The osmotic fragility of erythrocytes in rainbow trout under different dietary fatty acid status. *Fish. Sci.*, 1994 60, 93-95.

Kirpichnikov, VS. *Genetics and selection of fishes.* Leningrad: *Nauka* Publishers Leningrad branch; 1987.

Kirsipuu, A; Laugaste, K. About seasonal changes in the protein metabolism in the bream. In: Stroganov NS. *Contemporary problems of ecological physiology of the fishes.* Moscow: Nauka Publisher; 1979; 174-179.

Klotz, IM; Darnall, DW; Langerman, NR. Quaternary structure of proteins. In: Neurath H. *The Proteins.* New York: Academic Press.; 1975; 293-411.

Knoph, MB; Thorud, K. Toxicity of ammonia to Atlantic salmon(Samo salar L.)In seawater – effect on plasma osmolality, ion, ammonia and glucose levels and hematologic parameters. *Comp. Biochem. Physiol.*, 1966 113A, 375-381.

Kobayashi, K; Hara, A; Takano, K; Hirai, H. Studies on subunit components of immunoglobulin M from a bony fish, the chum salmon, Oncorhynchus keta. *Mol.Immun.*, 1982 19(1), 95-103.

Koehn, K. Serum haptoglobins in some American Catostomid fishes. *Comp.Biochem.Physiol.*, 1966 17(1), 349-352.

Kragh-Hansen, U. Structure and ligand binding properties of human serum albumin. *Dan. Med. Bull.*, 1990 37(1), 57-84.

Kragh-Hansen, U; Saito, S; Nishi, K; Anraku, M; Otagiri, M. Effect of genetic variation on the thermal stability of human serum albumin. *Biochim Biophys Acta*, 2005 1747, 81-88.

Kruse, Ch; Sordyl, H; Bestimmung der mittleren lebenszeit von erythrozytten bei regenbogenforellen (Salmo gairdneri R.) mittels Cr^{51}- markierung. *Wiss.Z.Wilhelm-Pieck-Univ., Rostock. Naturwiss.R.*, 1988 37, 93-96.

Kvingedal, AM; Rorvik, KA; Alestrom, P. Cloning and characterization of Atlantic salmon (Salmo salar) serum transferring cDNA. *Mol.Marine Biol.*, 1993 2(4), 233-238.

Kvingedal, AM. Characterization of the 5' region of the Atlantic salmon (Salmo salar) transferring-encoding gene. *Gene*, 1994 150(2), 335-339.

Kuzmin, EV. Allozyme Variation of Nonspecific Esterases in Russian Sturgeon (Acipenser guldenstadtii Brandt). Original Russian text is published in *Genetics*, 2002 38(4), 507-514.

Kuzmin, EV. Albumin system of the blood serum of Acipenseriformes in the river life cycle. Original Russian text is published in *Voprosy Ikhtiology*, 1996 36(1), 101-108.

Kuzmin, EV; Kuzmina, OY. Population Analysis of Electrophoretic Variation in Blood Serum Albumins of European (A. ruthenus L.) and Siberian (A. ruthenus marsiglii Brandt.) Sterlet. Original Russian text is published in *Genetics*, 2005 41(2), 246-253.

Lambert, MBT; Kelly, PP. The binding of phenylbutazone to bovine and horse serum albumin.*Rish Journal of Medical Science*, 1978 147(1), 193-196.

Landis, EM; Pappenheimer, JR. Exchange of substancesthrough the capillary walls. In: Handbook of Physiology. Circulation. Washington DC. *Am.Physiol.Soc.* 1963(2) II, 961-1034.

Lee, JA; Lee, HA; Sandler, PJ. Uraemia: is urea more important than we think? *Lancet*, 1991 338(8780),1438-1440.

Lichenstein, HS; Lyons, DE; Wurfel, MM; Johnson, DA; McGinley, MD; Leidli, JC; Trollinger, DB; Mayer, JP; Wright, SD; Zukowski, MM.

Afamin is a new member of the albumin, alpha-fetoprotein, and vitamin D-binding protein gene family. *J. Biol. Chem.*, 1994 269 (27), 18149-54.

Litman, GW; Frommel, D; Chartrand, SL; Finstad, L; Good, RA. Significance of heavy chain mass and antigenic relationship in immunoglobulin evolution. *Immunichem.*, 1971 8, 345-348.

Litman, GW; Rast, JP; Shamblott, MJ; Haire, RN; Michele, H; Roess, W; Litman, RT; Hinds-Frey, KR; Zilch, A; Amemiya, CT. Phylogenetic diversification of Lobb CJ, Olson M, Clem WL. Immunoglobulin light chain classes in teleost fish. *The Journal of immunology*, 1984 132, 1917-1923.

Lu, J; Stewart, AJ; Sadler, PJ; Pinheiro, TJ; Blindauer, CA. Albumin as a zinc carrier: properties of its high-affinity zinc-binding site. *Biochem.Soc.Trans.*, 2008 36, 1317-1321.

Ludwig, A; Belfiore, NM; Pitra, C; Svirsky, V; Jenneckens, I. Genome duplication events and functional reduction of ploidy levels in sturgeon (Acipenser, Huso and Scaphirhynchus). *Genetics*, 2001 158(3), 1203-1215.

Lukjanenko, BI; Habarov, MV. *Albumin system of the blood serum of the different in the ecology forms of sturgeon fishes.* Yaroslavl: UVD REA (Upper-Volga department of the Russian ecological academy); 2005.

Lundqvist, M; Bergten, E; Stromberg, S; Pilstrom, L. Ig light chain gene in the Siberian sturgeon (Acipenser baeri). *J.Immunol.*, 1996 157(5), 2031-8.

Lutz, PL; Robertson, JD. Osmotic constituents of the coelacanth Latimeria chalumnae Smith. *Biol.Bull.*, 1971 141, 553-560.

Maisey, JG. Chondrichthyan phylogeny: a look at the evidence. *J. Ven. Paleont.*, 1984 4, 359-371.

Manwell, C. The biology of myxine. In: Brodal A, Fange R. Norwegian Government Press.; 1963.

Marchalonis, J; Edelman, GM. Polypeptide chains of immunoglobulins from the smooth dogfish (Mustelus canis). *Science*, 1966 154(756), 1567-8.

Marchalonis, JJ; Schonfeld, SA. Polypeptide chain structure of sting ray immunoglobulin. *Biochim. et Biophys. Acta.* 1970.221(3), 603-611.

Martemjanov, VI. Ranges of the regulation of the concentration of sodium, potassium, calcium, magnesium in the plasma, the erythrocytes and the muscular tissue of the roach Rutilus rutilus in the nature conditions. Original Russian text is published in *Zhurnal Evolyutsionnoi biokhimii I fiziologii*, 2001 37(2),107-112.

Masseyeff, R; Godet, R; Gombert, J. Blood proteins of Protopterus annectens. Electrophoretic and immunoelectrophoretic study. *C.R. Seances Soc. Biol. Fil.*, 1963 157, 167-172.

Messager, JL; Stephan, G; Quentel, C; Baudin-Laurencin, F. Effects of dietary oxidized fish oil and antioxidant deficiency on histopathology, haematology, tissue and plasma biochemistry of sea bass Dicentrarchus labrax. *Aquat. Livig Resour. Vivantes Aquat.*, 1992 5, 205-214.

Mestel, R. Sharks mark a great divide in the evolution of immunity: before them, not a trace of antibodies or other pivotal immune proteins; after them, all elements are in place. *Natur. Hist.*, 1996 105(9), 40-44, 46.

Metcalf, V; Brennan, S; Chambers, G; George, P. The albumins of Chinook salmon (Oncorhynchus tshawytscha) and brown trout (Salmo trutta) appear to lack a propeptide. *Arch. Biochem. Biophys.*, 1998a 350(2), 239-244.

Metcalf, VJ; Brennan, SO; Chambers, GK; George, PM. The albumin of the brown trout (Salmo trutta) is a glycoprotein. *Biochim. Biophys. Acta,* 1998b 1386(1), 90-96.

Metcalf, VJ; Brennan, SO; Chambers, G; George, PM. High density lipoprotein (HDL), and not albumin, is the major palmitate binding protein in New Zealand long-finned (Anguilla dieffenbachii) and short-finned eel (Anguilla australis schmidtii) plasma. *Biochim. Biophys. Acta,* 1999a 1429(2), 467-475.

Metcalf, VJ; Brennan, SO; George, PM. The Antarctic toothfish (Dissostichus mawsoni) lacks plasma albumin and utilises high density lipoprotein as its major palmitate binding protein. *Comp. Biochem. Physiol., B., Biochem. Mol. Biol.,* 1999b 124(2), 147-155.

Metcalf, V; Brennan, St; Georg, P. Using serum albumin to inter vertebrate phylogenies. *Applied bioinformatics,* 2003 2(3), 97-107.

Metcalf, VJ; George, PM; Brennan, SO. Lungfish albumin is more similar to tetrapod than to teleost albumins: purification and characterization of albumin from Australian lungfish, Neocaratodus forsteri. Comp. biochemistry and physiology. Part B, *Biochemistry and molecular biology,* 2007 147(3), 428-437.

Mikrjakov, DV. Influence of some corticosteroid hormones to structure and function of the immune system of the fishes: dissertation abstract...candidate of biological science. Moscow; 2004; 1-24.

Mills, GL; Taylaur, CE; Chapman, MJ; Forster, GR. Characterization of serumlipoproteins of the shark Centrophorus squamous. *Biochem. J.,* 1977 163(3), 455-465.

Minchiotti, L; Campagnoli, M; Rossi, A; Cosulich, ME; Monti, M; Pucci, P; Kragh-Hansen, U; Granel, B; Disdier, P; Weiller, PJ; Galliano, M. A nucleotide insertion and frameshift cause albumin Kenitra, an extended and O-glycosylated mutant of human serum albumin with two additional disulfide bridges. *Eur. J. Biochem.*, 2001 268, 344-352.

Minchiotti, L; Galliano, M; Kragh-Hansen, U; Peters, TJr. Mutations and polymorphisms of the gene of the major human blood protein, serum albumin. *Hum.Mutat.*, 2008 29(8), 1007-1016.

Minghetti, PP; Ruffner, DE; Kuang, WJ; Dennison, OE; Hawkins, JW; Beattie, WG; Dugaiczyk, A. Molecular structure of the human albumin gene is revealed by nucleotide sequence within q11-22 of chromosome 4. *J. Biol. Chem.*, 1986 261(15), 6747-6757.

Mochida, K; Lou, YH; Hara, A; Yamauchi, K. Physical biochemical properties of IgM from a teleost fish. *Immunology*, 1994 83, 675-680.

Mouridsen, HT; Wallevik, K. Metabolic and gel-electrophoretic properties of albumin isolated from wound tissue. *Scand. J. Clin. and Lab.Invest.*, 1968 22(4), 322-327.

Mouridsen, HT. The retention of different plasma proteins in wound tissue. *Scand. J. Clin. and Lab. Invest.* 1969 23(3), 235-240.

Nakamura, O; Kudo, R; Aoki, H; Watanabe, T. IgM secretion and absorption in the materno-fetal interface of a viviparous teleost, Neoditrema ransonneti (Perciformes; Embiotocidae). *Dev.Comp.Immunol.*, 2006 30(5), 493-502.

Nash, AR; Tompson, EOP. Haemoglobin of the shark, Heterodontus portusjacksoni. *Austral. J. Biol. Sci.*, 1974 27(6), 607-615.

Nefedov, GN. Serum haptoglobins of the marine perch of the kind Sebastes. Original Russian text is published in *Vestnic Moskovskogo universiteta*, 1969 1, 104-108.

Nielsen, H; Kragh-Hansen, U; Minchiotti, L; Galliano, M; Brennan, SO; Tarnoky, AL; Franco, MH; Salzano, FM; Sugita, O. Effect of genetic variation on the fatty acid-binding properties of human serum albumin and proalbumin. *Biochim. Biophys. Acta*, 1997 1342, 191-204.

Novikov, GG. *Increase and power engineering of the development of bony fishes in the early ontogenesis.* Moscow: Editorial Publisher URSS; 2000.

Nunomura, W. C-reactive protein in eel: purification and agglutinating activity. *Biochim. Biophys. Acta,* 1991 1076(2), 191-196.

Obach, A; Quentel, C; Laurencin, FB. Effect of alpha-tocopherol and dietary oxidized fish oil on the immune response of sea bass // *Dis. Aquat. Org.* 1993. V.15. P.175-185.

Ohta, Y; Flajnik, M. IgD, like IgM, is a primordial immunoglobulin class perpetuated in most jawed vertebrates. *PNAS*, 2006 103(28), 10723-10728.

Olson, KR; Kinney, DW; Dombrowski, RA; Duff, DW. Transvascular and intravascular fluid transport in the rainbow trout: revisiting Starling's forces, the secondary circulation andinterstitial compliance. *J. Exp. Biol.*, 2003 206(3), 457-67.

Omura, Y; Korf, HW; Oksche, A. Vascular permeability (problem of the blood-brain barrier) in the pineal organ of the rainbow trout, Salmo gairdneri. *Cell Tissue Res.*, 1985 239(3), 599-610.

Ostensson, K; Lun, S. Transfer of immunoglobulins through the mammary endothelium and epithelium and in the local lymph node of cows during the initial response after intramammary challenge with E.coli endotoxin. *Acta Vet.Scand.*, 2008 50(1), 26.

Ota, T; Rast, JP; Litman, GW; Amemiya, CT. Lineage-restricted retention of a primitive immunoglobulin heavy chain isotype within the Dipnoi reveals an evolutionary paradox. *PNAS*, 2003 100(5), 2501-2506.

Pages, T; Gomez, E; Suner, O; Viscor, G; Tort, L. Effects of daily managements stress on haematology and blood rheology of the gilthead seabream // *J.Fish.Biol.* 1995. V.46. P.775-786.

Palmour, RM; Sutton, HE. Vertebrate transferrins molecular weight, clinical composition and iron binding studies // *Biochem.* 1971. V.10. P.4026-4032.

Pantjavin, AA; Artuhov, VG; Vashanov, GA. Modification of the physical chemistry properties of the molecules of serum albumin, induced by the vacuum ultraviolet radiation. Original Russian Text is published in *Vestnic VGU, series Chemistry, Biology*, 2000,122-125.

Papenfuss, HD; Gross, JF. Vasomotion and transvascular exchange of fluid and plasma proteins. *Microcirc.Endothelium Lymphatics.*, 1985 2(6), 577-96.

Papenfuss, HD; Hauck, G. The effect of the gradient of vascular permeability on fluid and plasma exchange in the mesenteric microcirculation. *Int.J.Microcirc.Clin.Exp.*, 1987 6(3), 203-213.

Papenfuss, HD; Gross, JF. Transvascular exchange of fluid and plasma proteins. *Biorheology*, 1987 24(3), 319-35.

Patterson, JE; Geller, DM. Bovine microsomal albumin: amino terminal sequence of bovine proalbumin.*Biochem.Biophys.Res.Commun.*, 1977 74, 1220-1226.

Peach, RJ; Brennan, SO. Structural characterization of a glycoprotein variant of human serum albumin: albumin Casebrook (494 Asp – Asp). *Biochim. Biophys Acta*, 1991 1097, 49-54.

Peters, MT; Davidson, LK. Isolation and properties of the fatty acidbinding protein from the Pacificc lamprey (Lampetra tridentate). *Comp.Biochem.Physiol.B.*, 1991 99, 619-623.

Phillips, MCL; Moyes, CD; Tufts, BL. The effects of cell ageing on metabolism in rainbow trout (Oncorhynchus mykiss) red blood cell. *J.Exp.Biol.*, 2000 203, 1039-1045.

Phillips, K. Trout with tone. *J.Exp.Biol.*, 2003 206, 424-426.

Piermarini, PM; Evans, DH. Osmoregulation of the Atlantic stingray (Dasyatis Sabina) from the freshwater lake Jesup of the St.Johns River, Florida. *Physiol.Zool.*, 1998 71, 553-560.

Pillans, RD; Franklin, CE. Plasma osmolyte concentrations and rectal gland mass of bull sharks Carcharhinus leucas, captured along a salinity gradient. *Comp.Biochem.Physiol. A Mol. Integr.Physiol.*, 2004 138(3), 363-371.

Polenov, SA; Dvoreckij, DP. Measurement of the trans-capillary exchange of the liquid. In: Methods of the study of the blood circulation. Leningrad: Nauka Publishers; 1976; 163-184.

Poljakovskij, VI; Pankovskaja, AA; Bogdanov, LV. Biochemical polymorphism of silver crucian, Carassius auratus gibelio Bl., dwelling in the lake Sudobl (Belarus). In: Biochemical genetics of the fishes. Leningrad: Nauka Publishers; 1973; 161-166.

Rall, D; Scchwab, P; Zubrod, Ch. Alteration of plasma proteins at metamorphosis in the Lamprey (Petromyzon marinus dosatus).*Science*, 1961 133(3448), 279-280.

Rast, JP; Anderson, MK; Ota, T; Litman, RT; Margittai, M; Shamblott, MJ; Litman, GW. Immunoglobulin light chain class multiplicity and alternative organizational form in early vertebrate phylogeny. *Immunogenetics*, 1994 40, 83-89.

Rippe, B; Haraldsson, B. Fluid and protein fluxes across small and large pores in the microvasculature. Application of two-pore equations. *Acta Physiol. Scand.*, 1987 131(3), 411-428.

Robey, FA; Tanaka, T; Liu, TY. Isolation and characterization of two major serum proteins from the dogfish, Mustelus canis, C-reactive protein and amyloid P component. *J. Biol. Chem.*, 1983 25(258/6), 3889-3894.

Roesner, A; Fuchs, C; Hankeln, T; Burmester, TA. 2005. Globin Gene of Ancient Evolutionary Origin in Lower Vertebrates: Evidence for Two

Distinct Globin Families in Animals. *Molecular Biology and Evolution*, 2005 22(1), 12-20.

Rogers, KA; Richardson, JP; Scinicariello, F; Attanasio, R. Molecular characterization of immunoglobulin D in mammals: immunoglobulin heavy constant delta genes in dogs, chimpanzees and four old world monkey species. *Immunology*, 2006 118(1), 88-100.

Romer, A; Parsons, T. *Anatomija pozvonochnih*. Moscow: Mir Publisher; 1992. Translated from Inglish.

Rumfelt, LL; Lohr, RL; Dooley, H; Flajnik, MF. Diversity and repertoire of IgW and IgM VH families in the newborn nurse shark. *J.BMC Immunol.*, 2004 5(1), 8.

Ryabtseva, IP; Mikrjakov, DV; Lukjanenko, VV; Andreeva, AM. Influence of dexamethasone and testosterone to resistant characteristics of erythrocyte of sterlet *ACIPENSER RUTHENUS*. Abstract book. Conference "Ecological problems of fresh water fish business reservoirs". Kazan, 2011.

Saber, MA; Stockbauer, P; Moravek, L; Meloun, B. Disulfide bonds in human serum albumin. *Collect.Czech.Chem.Commun.*, 1977 42, 564-579.

Saha, NR; Suetake, H; Suzuki, Y. Characterization and expression of the immunoglobulin light chain in the fugu: evidenceof a solitaire type // *Immunogenetics*. 2004. 56(1):47-55.

Saito, K. Etude biochimique du sang des Poissons. *Bull. Jap. Soc. Sci. Fish.*, 1957 22,752-759.

Sargent, TD; Yang, M; Bonner, J. Nucleotide sequence of cloned rat serum albumin messenger RNA. *Proc.Natl.Acad.Sci.U.S.A.*, 1981 78, 243-246.

Sarin, H. Physiologic upper limits of pore size of different blood capillary types and another perspective on the dual pore theory of microvascular permeability. *J. Angiogenes Res.*, 2010 2, 14.

Schmidt-Nielsen, K. Animal physiology. Cambridge University Press. Cambridge. 1979.

Schoentgen, F; Metz-Boutigue, MH; Jolles, J; Constans, J; Jolles, P. Complete amino acid sequence of human vitamin D-binding protein (group-specific component): evidence of a three-fold internal homology as in serum albumin and alpha-fetoprotein. *Biochim.Biophys.Acta*, 1986 871(2), 189-198.

Scholander, PF; Hargens, AR; Miller, SL.Negative pressure in the interstitial fluid of animals. *Science*, 1968 161(3839), 9-14.

Selje, G. The Story of The Adaptation Syndrome. Moscow: Government Medicine Publisher (MEDGIZ); 1960.

Serpunin, GG; Likhatchyova, OA. Use of the ichthyohaematological studies in ecological monitoring of the reservoirs. *Acta Vet.Brno*, 1998 67, 339-345.

Shamblott, MJ; Litman, GW. Genomic organization and sequences of immunoglobulin light chain genes in a primitive vertebrate suggest coevolution of immunoglobulin gene organization. *The EMBO Journal*, 1989 8, 3733-3739.

Shilov, IA. *Physiological ecology of the animals*. Moscow: Visshaja shkola (The higher school); 1985.

Shulman, GE. *Physiological biochemical special features of the annual cycles of the fishes*. Moscow: Food industry Publisher; 1972.

Shulz, GE; Schirmer, RH. Principles of protein structure. New York: Springer-Verlag New York Inc.; 1979.

De Smet, H; Blust, R; Moens, L. Absence of albumin in the plasma of the common carp Cyprinus carpio binding of fatty acids to high density lipoprotein. *Fish Physiology and Biochemistry*, 1998 19(1), 71-81.

Soldatov, AA. Special features of organization and functioning of the system of the red blood of the fishes. Original Russian text is published in *Zhurnal Evolyutsionnoi biokhimii I fiziologii*, 2005 41(3), 217-224.

Solem, ST; Jorgensen, TT. Characterisation of immunoglobulin light chain cDNAs of the Atlantic salmon, Salmo salar L.; evidence for three IgL isotypes. *Dev. Comp. Immunol.*, 2002 26(7), 635-47.

Sorkina, DA; Zalevskaja, IN. *Structural-functional properties of proteins*. Kiev: Higher school Publisher; 1989.

Speers-Roesch, B; Ip, YK; Ballantyne, JS. Metabolic organization of freshwater, euryhaline, and marine elasmobranches: implication for the evolution of energy metabolism in shaks and rays. *J. Exp. Biol.*, 2006 209(13), 2495-2508.

Starling, EH. On the absorption of fluids from the connective tissue spaces. *J. Physiol. (London)*, 1895 19, 312-326.

Strauss, AW; Bennett, CD; Donohue, AM; Rodkey, JA; Alberts, AW. Rat liver pre-proalbumin: complete amino acid sequence of the pre-piece. Analysis of the direct translation product of albumin messenger RNA. *J.Biol.Chem.*, 1977 252, 6846-6855.

Stroganov, NS. Ecological physiology of the fishes. Moscow: Moscow State University, MSU; 1962.

Subbotkin, MF; Subbotkina, TA. Comparative immunochemical analysis of the antigens of the serum proteins of large Amu-Darya shovelnose sturgeon, Pseudoscaphirhynchus Kaufmanni (Acipenseridae). Original Russian text is published in *Voprosy Ikhtiology*, 2003 43(2), 254-261.

Subbotkin, MF; Subbotkina, TA. Age-related variability of serum protein antigenns in sturgeons (Acipenseridae, Acipenserifomes). Original Russian text is published in *Ontogenez*, 2004 35(5), 360-366.

Subbotkin, MF; Subbotkina, TA. Ontogenetic variation of serum protein antigens of the large Amu-Darya shovelnose sturgeon, Pseudoscaphirhynchus Kaufmanni (Acipenseridae). *J. App. Ichthyol.*, 2011 27, 213-218.

Sugio, S; Kashima, A; Mochizuki, S; Noda, M; Kobayashi, K. Crystal structure of human serum albumin at 2.5A resolution. *Protein Eng.*, 1999 12(6), 439-46.

Sulya, LL; Box, BE; Gordon, G. Plasma proteins in the blood of fishes from the Gulf of Mexico. *Am. J. Physiol.*, 1961 200, 152-154.

Szebedinszky, C; Gilmour, KM. The buffering power of plasma in brown bullhead (Ameiurus nebulosus). *Comp. Biochem. Physiol., B., Biochem. Mol. Biol.*, 2002 131(2), 171-183.

Takahashi, N; Takahashi, Y; Putnam, FW. Complete amino acid sequence of human hemopexin, the heme-binding protein of serum. *Proc.Natl.Acad.Sci. U.S.A.* 1985 82(1), 73-7.

Terskov, IA; Gitelzon, II. Method of the chemical acidic erythrograms. *Biophysics*, 1957 2, 167-172.

Thakurta, PG; Choudhury, D; Dasgupta, R; Dattagupta, JK. Tetriary structural changes associated with iron binding and release in hen serum transferring: a crystallographic and spectroscopic study. *Biochem.Biophys.Res.Commun.*, 2004 316(4), 1124-31.

Thompson, JG; Simpson, AC; Pugh, PA; Tervit, HR. In vitro development of early sheep embryos is superior in medium supplemented with human serum compared with sheep serum or human serum albumin. *Animal Reproduction Science.* 1992 29(1), 61-68.

Thorson, TB; Cowan, CM; Watson, DE. Body fluid solutes of juveniles and adults of the euryhaline bull shark Carcharinus leucas from freshwater and saline environments. *Physiol.Zool.*, 1973 46, 29-42.

Tinaeva, EA; Markovich, LG; Konkina, VV; Semikrasova, EA. About possibility of blood proteins polymorphism using as the index of selection in fur farming. *Vestnik VOG i C* (Russia), 2007 11(1), 122-130.

Tsvetnenko, YB. Special features of capillary permeability for the endogenous proteins and the lipids of plasma in the fishes. In: Abstract Book of 9th Confer. *"Problems of the evolutionary physiology"*. Leningrad:Institute of evolutionary physiology and biochemistry;1986; 304-305.

Valenta, M; Stratil, A; Slechtova, V; Kalal, L. Polymorphism of transferring in carp (Cyprinus carpio L.) – genetic determination, isolation and partial characterization. *Biochem.Genet.*, 1976 14(1-2), 27-45.

Vazquez-Moreno, L; Porath, J; Schluter, SF; Marchalonis, JJ. Purification of a novel heterodimer from shark (Carcharhinus Plumbeus) serum by gel-immobilized metal chromatography. *Comp. Biochem. Physiol. B.*, 1992 103(3), 563-568.

Villalobos, AR; Renfro, JL. Trimethylamine oxide suppresses stress-induced alteration of organic anion transport in choroids plexus. *J.Exp.Biol.*, 2007 210(3), 541-552.

Vollf, JN. Genome evolution and biodiversity in teleost fish. *Heredity*, 2005 94(3), 280-294.

Wally, J; Halbrooks, PJ; Vonrheim, C; Rould, MA; Everse, SJ; Mason, AB; Buchanan, SK. The crustal structure of iron-free human serum transferring provides insight into inter-lobe communication and receptor binding.*J.Biol.Chem.*, 2006 281(34), 24934-24944.

Wally, J; Buchanan, SK. A structural comparison of human serum transferring and human lactoferrin. *Biometals*, 2007 20(3-4), 249-62.

Walsh, P; Wood, C; Perry, S; Thomas, S. Urea transport by hepatocytes and red blood cells of selrcted elasmobranch and teleost fishes. *J. Exp. Biol.*, 1994 193(1), 321-35.

Wang, X; Wu, L; Aouffen, M; Mateescu, MA; Nadeau, R; Wang, R. Novel cardiac protectiveeffects of urea: from shark to rat. *Br. J. Pharmacol.*, 1999 128(7), 1477-1484.

Wang, D; Liu, HB. Immunoglobulin heavy chain constant region of five Acipenseridae: cDNA sequence and evolutionary relationship. *FishShellfish Immunol.*, 2007 23(1), 46-51.

Watkins, S; Madison, J; Galliano, M; Minchiotti, L; Putman, FW. Analbuminemia: three cases resulting from different point mutations in the albumin gene. *Proc. Natl. Acad. Sci. U.S.A.* 1994 91, 9417-9421.

Weber, RE; Vinogradov, SN. Nonvertebrate hemoglobins: functions and molecular adaptations. *Physiol.Rev.*, 2001 81, 569-628.

Wendelaar, Bonga; Sjoerd, E. The stress response in fish. *Physiol. Rev.*, 1997 77(3), 591–625.

White, A; Handler, Ph; Smith, E; Hill, R; Lehman, I. Principles of biochemistry. New York: McGraw-Hill Inc. 6th ed.; 1978.

Wilson, M; Bengten, E; Miller, NW; Clem, LW; Pasquer, L; Warr, GW. A novel chimeric Ig heavy chain from a teleost fish shares similarities to IgD. *Proc. Natl. Acad. Sci. USA*, 1997 94, 4593-4597.

Woodbury, RG; Brown, JP; Yeh, MY; Hellstrom, I; Hellstrom, KE. Identification of a cell surface protein, p97, in human melanomas and certain other neoplasms. *Proc. Natl. Acad. Sci. U.S.A.* 1980 77, 2183-2187.

Wuertz, S; Nitsche, A; Jastroch, M; Gessner, J; Klingennspor, M; Kirschbaum, F; Kloas, W. The role of the IGF-I system for vitellogenesis in maturing female starlet, Acipenser ruthenus Linnaeus, 1758. *Gen.Comp.Endocrinol.*, 2007 150(1), 140-150.

Xu, Y; Ding, Z. N-Terminal Sequence and Main Characteristics of Atlantic Salmon (Salmo salar) *Albumin. Prep. Biochem. Biotechnol.*, 2005 35(4), 283-290.

Yakovlev, VN. Phylogenesis of Acipenseriformes. In: Stories about phylogeny and systematics of the fossil fishes and jawless. Moscow: *Nauka* Publisher; AS USSR; 1977; 116-144.

Yasuike, M; Boer, J; Dchalburg, K; Cooper, G; McKinnel, L; Messmer, A; So, S; Davidson, W; Koop, B. Evolution of duplicated*IgH* loci in Atlantic salmon*Salmo salar.BMC Genomics*, 2010 11, 486.

Zhou, Xian-Qing; Sun, Ru-Yong; Niu, Cui-Juan. Influence of stress on the growth, behavior and physiological activity of water animals. *Dongwuxue yanjiu (Zool. Res.)*, 2001 22(2), 154-158.

Zorin, NA; Gabin, SG; Likova, OF; Zorina, VN; Belogorlova, TI; Chirikova, TS. Comparative study of the physical chemistry and antigenic properties of albumins of human and animals.Original Russian text is published in *Journal of Evolutionary biochemistry and physiology,* 1994 30(4), 505-511.

Zweifach, BW; Intaglietta, M. Mechanics of fluid movement across single capillaries in the rabbit. *Microvasc.Res.,* 1968 I, 83-101.

INDEX

A

acid, 9, 11, 14, 17, 21, 27, 28, 29, 31, 32, 33, 44, 47, 48, 50, 52, 53, 59, 62, 65, 66, 67, 68, 84, 110, 111, 115, 116, 119, 123, 148, 166, 169, 173
acidic, 122, 123, 124, 130, 131, 133, 178
Acipenseriformes, 41, 42, 44, 45, 46, 48, 63, 137, 144, 148, 150, 153, 154, 155, 170, 180
active compound, 99
adaptability, 49
adaptation, 94, 97, 133, 146, 151, 153, 156
adaptations, 71, 107, 155, 156, 179
adults, 178
age, 87, 128, 131, 134
aggregation, 37, 38, 46, 145
2, 173, 174, 175, 176, 177, 178, 179
albumin Kenitra, 10, 173
algorithm, 107, 156
alpha-fetoprotein, 9, 167, 171, 177
alpha-tocopherol, 174
amino, 8, 9, 11, 16, 17, 18, 20, 22, 27, 32, 34, 44, 48, 50, 51, 58, 59, 65, 67, 74, 84, 99, 104, 105, 110, 115, 116, 119, 148, 149, 164, 166, 175, 176, 177, 178
amino acid, 8, 9, 16, 17, 18, 20, 22, 27, 32, 34, 51, 58, 65, 74, 99, 104, 105, 116, 148, 149, 164, 166, 176, 177, 178
amino acids, 8, 9, 16, 17, 18, 20, 22, 27, 32, 34, 51, 65, 74, 99, 104, 105, 116, 148, 166
ammonia, 170
ammonium, 5, 43, 48, 113, 114, 115, 119, 120, 121, 122, 134, 135, 160
amphibians, 23, 65
antigen, 27
antioxidant, 172
apoptosis, 133, 140, 150
aquarium, 95, 99, 100, 131, 133
Asia, 163
astrocytes, 91
autonomy, 136

B

bacillus, 43
base, 101, 113, 114, 115
Belarus, 175
biochemistry, 160, 161, 164, 168, 172, 179, 180
biodiversity, 179
bioinformatics, 172
biological fluids, 156
biological processes, 169
birds, 67
birefringence, 43
blood circulation, 5, 175
blood flow, 104, 135, 138, 139, 150

blood smear, 132
blood stream, 17
blood vessels, 5, 7, 8, 91, 164
blood-brain barrier, 169, 174
bloodstream, 2, 4
bonds, 9, 10, 15, 16, 17, 19, 20, 21, 22, 23, 38, 53, 67, 75, 85, 103, 107, 138, 176
brain, 8, 56, 79, 87, 88, 89, 90, 91, 92, 93, 96
breeding, 165
Brno, 177

C

Ca^{2+}, 11, 34
calcium, 171
capillary, 1, 5, 6, 7, 8, 24, 87, 90, 91, 92, 93, 99, 106, 107, 146, 147, 151, 152, 154, 155, 166, 168, 170, 175, 176, 179
carbohydrate, 27, 48, 72, 102, 141, 156, 157
carbohydrates, 8, 10, 16, 34, 45, 72, 104, 141
Carcharhinus plumbeus, 28, 34, 38
cartilaginous, ix, 2, 26, 27, 32, 37, 38, 39, 41, 43, 45, 53, 110, 137, 141, 142, 143, 144, 145, 146, 147, 150, 151, 153, 154, 155, 157, 160
Caspian Sea, 41, 46
catabolism, 63, 84
catfish, 51, 55, 56, 67, 163, 167
cDNA, 148, 159, 164, 167, 168, 170, 179
cell surface, 15, 180
changing environment, 94
chemical, 7, 160, 178
chemical properties, 160
chromatography, 42, 113, 179
chromosome, 9, 18, 20, 22, 23, 24, 51, 59, 60, 63, 116, 117, 173
circulation, 174
classes, 43, 55, 153, 171
cleaning, 140
cleavage, 57, 74, 83, 84, 148, 163
clone, 29, 166, 167, 168
clusters, 16, 23, 27, 154
coding, 23, 163

codominant, 11, 50
codon, 163
color, 4, 43
communication, 179
community, 153
complement, 26, 163
compliance, 174
composition, 1, 2, 16, 26, 36, 43, 44, 48, 51, 58, 60, 61, 67, 71, 75, 77, 78, 85, 92, 102, 103, 104, 109, 112, 132, 135, 145, 146, 147, 152, 153, 154, 155, 167, 174
compounds, 26, 91
conception, 7
conference, 160, 162, 166
configuration, 155
connective tissue, 177
constituents, 171
control group, 95, 123
correlation, 45, 92, 99, 166
cortisol, 166
covalent bond, 8, 34, 53
C-reactive protein, 34, 38, 68, 173, 176
crystal structure, 169
crystals, 114, 115, 150
culture, 63
cycles, 177
cytochrome, 150
cytochromes, 150
cytology, 165
cytoplasm, 136

D

database, 56, 69, 84, 102
decomposition, 119, 139, 150, 152
deficiency, 172
deficit, 109
degradation, 2, 3, 74, 75, 76, 104, 105, 109, 132, 138, 139, 150
dehydration, 113, 119, 120, 121, 122, 133, 135, 136, 155
destruction, 2, 113, 115, 119, 134, 138, 139, 140, 150, 151
destruction processes, 134
detection, 37, 60, 64, 65, 84, 107, 146

Index

diabetes, 10, 154
diet, 163
dietary fat, 169
diffusion, 7, 26
dimerization, 166
diploid, 142
disorder, 11
displacement, 106, 107
dissociation, 24, 36, 37, 38, 46, 103, 106, 107, 145, 146, 152, 156, 157
distilled water, 119, 127, 128, 130
distribution, 74, 93, 155, 161, 164
diversification, 171
diversity, ix, 1, 2, 28, 30, 159, 162, 168
DNA, 132, 133, 150, 164
dogs, 176
drainage, 4
dyeing, 72
dyes, 85, 102

E

E.coli, 174
ecology, 162, 171, 177
electron, 167
electrophoresis, 11, 30, 31, 35, 36, 37, 42, 44, 45, 46, 47, 50, 53, 54, 55, 61, 64, 68, 69, 70, 71, 72, 73, 76, 77, 78, 79, 80, 81, 82, 89, 94, 102, 103, 111, 112, 113, 114, 119, 132, 133, 137, 143, 154
emission, 140
encoding, 16, 116, 167, 170
endothelial cells, 7, 91
endothelium, 163, 174
energy, 99, 105, 177
energy supply, 99
engineering, 173
environment, 1, 3, 26, 126, 135, 140, 147, 151, 152, 153, 156, 162
environmental conditions, 153
environmental factors, 87, 150
enzyme, 43
epithelium, 174
equilibrium, 93
erythrocyte membranes, 131
erythrocytes, 2, 17, 59, 109, 119, 122, 123, 124, 125, 126, 127, 128, 129, 130, 131, 132, 133, 134, 135, 136, 139, 150, 151, 169, 171
estrogen, 166
ethanol, 48, 54
evidence, 11, 32, 68, 159, 171, 176, 177
evolution, 1, 2, 16, 44, 46, 48, 148, 149, 150, 151, 153, 154, 155, 164, 167, 168, 171, 172, 177, 179
excretion, 22, 39, 164
experimental condition, 146
external environment, 2

F

families, 41, 56, 152, 153, 160, 167, 176
fast-fraction heterogeneity, 36
fat, 11
fatty acids, 68, 177
fertility, 49
fibrinogen, 4
filtration, 2, 5, 90, 91, 92, 93, 94, 95, 99, 156, 159, 162
fish oil, 172, 174
fluctuations, 25, 39, 88, 95, 135, 136
fluid, 1, 2, 3, 4, 5, 6, 7, 8, 26, 27, 30, 35, 36, 37, 56, 79, 87, 88, 89, 90, 91, 92, 93, 94, 95, 97, 98, 99, 105, 106, 107, 140, 147, 152, 155, 156, 160, 162, 163, 164, 174, 177, 178, 180
fluid balance, 5, 8, 100
food, 99, 105, 131, 132, 156
force, 5
formation, 2, 3, 11, 20, 42, 43, 80, 113, 115, 135, 153, 155, 156, 164
formula, 16, 17, 19, 20
fractional composition, 88, 164
fragility, 169
fragments, 34, 57, 65, 67, 74, 83, 84, 140, 148
free volume, 54
freezing, 113, 115, 135, 150
freshwater, 41, 46, 162, 175, 177, 178

G

gamma globulin, 53
gel, 113, 133, 173, 179
gene expression, 3
genes, 4, 16, 22, 23, 24, 27, 28, 44, 55, 84, 110, 115, 116, 117, 118, 148, 150, 154, 157, 161, 162, 165, 167, 176, 177
genetics, 165, 175
genome, 16, 17, 28, 46, 50, 157
genus, 59
Germany, 69
gland, 164, 169, 175
global climate change, 1
gluconeogenesis, 99, 105
glucose, 99, 105, 109, 170
glycoproteins, 44, 51, 141
glycosylation, 10, 142
gonads, 63, 73, 84, 101, 104, 156
grass, 50, 59
growth, 166, 180
Gulf of Mexico, 178

H

habitat, 155
haptoglobin, 5, 17, 18, 19, 20, 32, 33, 42, 49, 59, 60, 61, 109, 137, 138, 139, 150, 153, 154, 167
HE, 174
health, 162
heme, 22, 109, 111, 112, 119, 135, 138, 150, 178
hemoglobin, 2, 3, 5, 17, 22, 23, 59, 60, 109, 111, 112, 113, 114, 115, 116, 117, 118, 119, 120, 121, 124, 127, 128, 129, 132, 133, 134, 135, 137, 138, 139, 140, 150, 151, 152, 155, 160, 161
hemorrhage, 165
hepatocytes, 63, 179
heterogeneity, 11, 27, 36, 42, 43, 50, 72, 101, 121, 142, 160
high density lipoprotein, 172, 177
histidine, 67

homeostasis, 3, 133, 135, 161
hormones, 11, 122, 124, 172
human, 10, 11, 16, 17, 18, 19, 20, 21, 22, 23, 27, 38, 46, 51, 60, 80, 148, 149, 154, 159, 163, 166, 168, 169, 170, 173, 175, 176, 178, 179, 180
hybridization, 164, 165
hydrocortisone, 122
hydrogen, 37, 38
hydrogen bonds, 37, 38
hypothesis, 5
hypovolemia, 88

I

identification, 46, 64
identity, 27, 48, 56, 69, 148
immune response, 154, 174
immune system, 166, 172
immunity, 159, 172
immunoelectrophoresis, 68
immunoglobulin, 15, 16, 27, 28, 29, 51, 58, 154, 159, 165, 167, 168, 170, 171, 174, 176, 177
immunoglobulins, 5, 15, 16, 27, 30, 37, 42, 43, 49, 51, 53, 55, 56, 144, 145, 147, 153, 154, 157, 162, 165, 171, 174
immunoreactivity, 167
in vitro, 37, 134, 145, 147
in vivo, 10, 17, 159
individuals, 46, 95, 96, 127, 129
industry, 177
insertion, 173
insulin, 163
integration, 153, 156
integrity, 109
interface, 173
intermolecular interactions, 142
internal environment, 2, 3, 41, 107, 155, 156
internal fluid, 1, 2, 3, 26, 39, 136, 140, 144, 145, 146, 147, 152, 156, 162
interstitial fluid, 2, 3, 4, 5, 7, 8, 35, 36, 37, 87, 88, 89, 90, 91, 92, 93, 94, 95, 96, 97, 98, 99, 105, 106, 107, 164, 177
introns, 23

iron, 2, 20, 22, 109, 136, 137, 138, 139, 140, 141, 150, 151, 152, 154, 159, 168, 174, 178, 179
isolation, 179

J

Japan, 69, 77, 130
juveniles, 178

K

K^+, 11
kidneys, 164

L

lactate dehydrogenase, 161
lactoferrin, 20, 179
larva, 50
lead, 11, 115, 157
life cycle, 132, 148, 165, 166, 170
ligand, 169, 170
light, 15, 16, 134, 159, 165, 167, 168, 171, 175, 176, 177
lipids, 34, 45, 109, 163, 179
lipoproteins, 37, 72, 173
liver, 8, 32, 39, 90, 164, 177
localization, 3, 8, 22
loci, 23, 28, 50, 51, 55, 56, 71, 165, 180
locus, 11, 28, 42, 44, 50, 72, 168
low-molecular proteins, 31, 34, 35, 36, 37, 38, 39, 70, 71, 72, 73, 78, 81, 88, 94, 96, 97, 99, 101, 105, 106, 107, 138, 141, 142, 145, 146, 147, 154
lungfishes, 48
lymph, 3, 4, 174
lymph node, 174
lymphatic system, 4
lymphocytes, 15, 16
lysozyme, 166

M

macromolecules, 91
magnesium, 171
majority, 11, 25, 42, 43, 48, 49, 51, 81, 85, 126, 148
mammal, 11, 65, 154
mammals, ix, 2, 3, 11, 15, 20, 27, 37, 38, 41, 44, 46, 48, 53, 65, 67, 91, 141, 142, 147, 148, 151, 153, 154, 157, 176
marine fish, 77, 80, 81, 84, 104, 119, 121, 126, 136, 152, 156
marine species, 88, 156
mass, 56, 69, 74, 82, 128, 132, 139, 171, 175
matrix, 168
media, 136
melanoma, 180
membranes, 15, 124, 128, 130, 136
mesentery, 90
messenger RNA, 176, 177
Metabolic, 173, 177
metabolism, 1, 45, 105, 107, 132, 136, 142, 147, 151, 153, 156, 162, 167, 169, 175, 177
metabolites, 8, 11
metal ion, 11
metal ions, 11
metals, 11
metamorphosis, 175
microcirculation, 1, 174
microscope, 127
migration, 156
mineralization, 135
models, ix, 1, 2, 3, 147, 148
modifications, 17, 84
molecular biology, 172
molecular weight, 1, 8, 14, 43, 91, 92, 114, 162, 174
molecules, 2, 9, 11, 38, 44, 53, 59, 60, 91, 99, 157, 174
monomers, 1, 11, 14, 24, 30, 37, 38, 72, 115, 135, 145, 147, 154, 155
Moscow, 159, 161, 166, 169, 172, 173, 176, 177, 178, 180

motor activity, 133
mRNA, 63, 166
muscles, 36, 56, 79, 87, 88, 90, 92, 93, 94, 97, 98, 105
muscular tissue, 171
mutant, 10, 173
mutation, 10, 167
mutational analysis, 164
mutations, 11
myoglobin, 22, 79, 82, 150

N

Na^+, 11, 39
NaCl, 126, 127, 128, 129, 130, 132, 134, 145
nervous system, 164
New Zealand, 172
NH2, 27, 34, 48, 65, 67, 68, 148
nickel, 11, 65
Nile, 53
nucleotide sequence, 165, 173

O

oligomeric complexes, 53, 145, 157
oligomerization, 37, 142, 155
oligomers, 1, 24, 81, 85, 145, 146, 147, 154, 155
organ, 91
organism, 2, 3, 4, 5, 20, 24, 26, 41, 56, 74, 94, 99, 105, 133, 135, 136, 140, 145, 151, 152, 153, 154, 155, 156, 161
osmolality, 170
osmotic pressure, 24, 26, 39, 152
ox, 120
oxidation, 109
oxyhemoglobin, 113, 114, 128, 130, 135

P

PAGE gradient, 35, 36, 44, 54, 70, 81, 114
pathology, 14, 154, 162
PCR, 163

penalties, 168
peptidase, 17, 163
peptides, 148
peritoneum, 87, 92
permeability, 8, 87, 90, 91, 92, 93, 99, 146, 147, 148, 151, 152, 155, 157, 161, 164, 168, 169, 174, 176, 179
peroxide, 109
pH, 54, 109
phenotype, 7, 46
phenotypes, 46
phosphate, 109, 122
physical chemistry, 174, 180
Physiological, 167, 177
physiology, 160, 161, 164, 169, 172, 176, 178, 179, 180
pineal organ, 91, 174
plasma proteins, 1, 2, 3, 4, 5, 7, 8, 41, 80, 82, 87, 90, 91, 92, 93, 94, 95, 99, 105, 107, 140, 141, 142, 146, 147, 151, 153, 155, 161, 173, 174, 175
plasticity, 168
plexus, 179
ploidy, 171
PM, 163, 172, 175
point mutation, 179
polar, 151
polymerization, 112, 166
polymorphism, 163, 175, 179
polymorphisms, 173
polypeptide, 1, 8, 10, 17, 19, 22, 32, 81, 154, 155, 156
polyploid, 50, 142
ponds, 87, 128
population, 46, 165
potassium, 171
precipitation, 48, 121, 134, 135
predators, 136
preparation, 130
principles, 3, 141
probe, 43
protection, 162
protein bonds, 142
protein sequence, 14
protein structure, 177

proteolysis, 75, 76, 84, 99, 104, 149
pumps, 156
purification, 172, 173

Q

Queensland, 148

R

radiation, 174
radio, 164
REA, 171
reaction medium, 38, 145
reading, 10
receptors, 15
recombination, 17, 154
reconstruction, 107
red blood cells, 179
redistribution, 95, 97, 99, 104, 146, 156
regeneration, 169
reproduction, 165
residues, 11, 17, 32, 44, 50, 59, 67, 110, 115, 116, 119, 156, 157
resistance, 109, 121, 122, 123, 124, 126, 127, 128, 129, 130, 131, 132, 133, 134, 135, 136, 139, 151
resolution, 9, 168, 169, 178
resources, 162, 166
respiration, 110
response, 7, 166, 174, 180
restoration, 60, 88
rheology, 174
Russia, 41, 166, 179

S

saline water, 69, 97, 98, 99
salinity, 1, 2, 25, 39, 41, 45, 49, 71, 95, 97, 99, 105, 106, 107, 126, 136, 146, 152, 155, 156, 169, 175
salmon, 51, 56, 64, 65, 66, 67, 119, 148, 149, 164, 170, 172, 177, 180
salt concentration, 26, 121, 145

salts, 41, 49, 144, 145
saturation, 113, 114, 115, 120, 121, 134
school, 177
science, 160, 161, 172
seasonal changes, 169
secretion, 173
sediment, 126
sedimentation, 16
selectivity, 151
sensitivity, 168
serum albumin, 9, 10, 14, 45, 46, 67, 70, 83, 85, 102, 103, 104, 141, 154, 159, 160, 162, 163, 164, 166, 168, 169, 170, 172, 173, 174, 175, 176, 177, 178
serum transferrin, 20, 21, 160, 170, 178, 179
sex, 131
sheep, 11, 164, 178
shock, 123
shortage, 156
sialic acid, 42, 43, 50, 65, 142, 160
Siberia, 165
signs, 44, 122
silver, 50, 75, 102, 124, 175
single cap, 180
sodium, 48, 163, 171
solubility, 5, 37, 48
solution, 5, 37, 43, 113, 114, 115, 119, 120, 123, 124, 125, 126, 128, 130
specialization, 2, 140, 151, 155
species, 1, 25, 27, 37, 38, 41, 42, 48, 49, 50, 51, 59, 63, 67, 69, 81, 85, 90, 91, 100, 112, 119, 120, 124, 126, 130, 131, 132, 136, 139, 143, 145, 146, 147, 155, 156, 160, 163, 164, 165, 166, 176
spleen, 167
stability, 2, 110, 119, 121, 124, 129, 130, 131, 133, 134, 135, 136, 150, 160
stabilization, 1, 2, 104, 107, 110, 135, 152, 156, 162
starvation, 87, 99, 132
state, 5, 87, 151, 153, 162, 165
storage, 119, 120, 121, 122, 124, 132, 135
stress, 7, 122, 124, 163, 174, 179, 180
stress factors, 124
structural changes, 178

structural transformations, 2, 39, 104, 105, 107, 146, 152, 156
structure, 1, 9, 10, 11, 16, 20, 21, 22, 27, 28, 30, 39, 44, 48, 55, 56, 59, 60, 65, 67, 72, 104, 141, 148, 154, 157, 159, 160, 162, 164, 168, 170, 171, 172, 173, 178, 179
subdomains, 9
substitutions, 163
sulfate, 43, 48, 113, 114, 115, 119, 120, 121, 122, 135, 160
Sun, 180
suppliers, 105
surface structure, 8, 37, 38, 85, 142, 148, 154, 156, 157
surplus, 138
survival, 139
susceptibility, 134
Sweden, 164
synthesis, 63, 105

T

taxa, 2, 136, 143, 144, 147, 148, 150, 153, 154
taxonomy, 153
temperature, 45, 105, 109, 128, 132
territory, 41
testosterone, 123, 124, 176
tetrapod, 172
thermal stability, 170
tissue, 4, 5, 6, 7, 22, 27, 56, 79, 82, 87, 88, 89, 97, 98, 105, 107, 145, 147, 148, 155, 162, 163, 164, 172, 173
tracks, 79, 124
transcapillary exchange, ix, 1, 2, 8
transferrin, 5, 20, 30, 42, 43, 49, 50, 51, 52, 95, 96, 109, 137, 138, 139, 140, 150, 154, 165
transformations, 37, 39, 101, 102, 107, 153
translation, 177

transport, 1, 5, 7, 11, 45, 106, 163, 165, 174, 179
treatment, 43
twins, 114

U

ultrasound, 130
untranslated regions, 23
urea, 26, 27, 30, 34, 36, 37, 38, 39, 41, 46, 47, 49, 60, 75, 76, 77, 78, 79, 81, 82, 89, 102, 103, 113, 114, 115, 135, 143, 144, 145, 147, 155, 163, 167, 171, 179
urine, 22, 24
USA, 166, 180
USSR, 160, 163, 169, 180

V

vacuum, 174
vascular wall, 3, 91
vertebrates, 1, 4, 34, 38, 42, 48, 49, 52, 59, 70, 90, 91, 109, 150, 153, 154, 166, 174
vessels, 4, 6, 7, 151
viscosity, 8
vitamin D, 9, 167, 171, 176

W

Washington, 170

Y

yolk, 4

Z

zinc, 67, 163, 171